MW01677016

Late Quaternary Environments and Deep History:

A Tribute to Paul S. Martin

Edited by
David W. Steadman and Jim I. Mead

Late Quaternary Environments and Deep History: A Tribute to Paul S. Martin

Edited by

David W. Steadman
New York State Museum

Jim I. Mead
Northern Arizona University

The Mammoth Site of Hot Springs, South Dakota, Inc.
Scientific Papers, Volume 3

Hot Springs, South Dakota
January 1995

LATE QUATERNARY ENVIRONMENTS AND DEEP HISTORY: A TRIBUTE TO PAUL S. MARTIN produced by The Mammoth Site of Hot Springs, South Dakota, Inc., 1800 Highway 18-Truck Route, Hot Springs, South Dakota 57747-0606. (605) 745-6017 or FAX (605) 745-3038

Printed in Rapid City, South Dakota by Fenske Printing Inc.

ISBN 09624750-4-1

The picture on the cover is the Desert Laboratory (Tumamoc Hill), west of downtown Tucson, Arizona. Built in 1903 for use by the Carnegie Institution, Washington, it has been the focus of intensive paleoecological research. Paul Martin has been a prominent figure on 'The Hill' where he has provided an atmosphere of scholarship for many students of the Pleistocene. Photo by Pete Kresan. Back cover—Paul Martin at fumarole, Isla Santa Fé, Galápagos Islands, January 1981. Photo by David W. Steadman.

Contents

Preface

It really makes no difference whether or not you agree with Paul Martin's ideas, models, and theories - if you conduct research about Pleistocene paleoecology, paleontology, or archaeology, you have been affected by Paul S. Martin. He is innovative, provacative, and incredibly subtle at being demanding. Few scholars have provided such stimulation for world-wide research and student growth as Paul.

Although 'retired', Paul is still at the Desert Laboratory, University of Arizona, Tucson. We wanted to have a symposium in tribute to Paul, so we asked him to produce a list of people that he felt should be present at such an event. From that lengthy register we finalized some 17 participants who could attend the tribute, which was organized by Larry Agenbroad and Jim Mead. From September 23 through 25, 1993, The Mammoth Site, Hot Springs, South Dakota, hosted the symposium *Late Quaternary Environments and Deep History (A Tribute to Paul S. Martin)*. Most of the presented talks are published here in the symposium volume, edited by Dave Steadman and Jim Mead. In addition to the authors in this volume, Valentina Ukraintseva, Tom Stafford, Joan Coltrain, Julio Betancourt, and Peter Mehringer also spoke at the symposium in Hot Springs.

Many of us were students of Paul while working for various degrees at the University of Arizona, Department of Geosciences. Paul also has affected others more distant from the clutches of an academic committee. His controversial ideas and paradigms stretch around the world. The themes on the pages that follow are as diverse as is Paul. He insists that you think carefully about whichever research topic grabs your interest. He has 'this way' of making you approach your problem from multiple directions—just to make sure that your thoughts are not unduly channeled. Although many of his students and near-students gleaned a research topic or technique from Paul, much more importantly, we all learned to take a multidisciplinary approach to our studies.

David W. Steadman
Jim I. Mead
January, 1995

Chapter 1

Rediscovering the Desert Lab

PAUL S. MARTIN

Emeritus Professor of Geosciences
Desert Laboratory, Department of Geosciences, University of Arizona, Tucson, Arizona 85721

The tradition of an academic festschrift ("feast writing") by former students and colleagues in honor of some patriarchal "Herr Doctor Professor" (do Frau Doctor Professors succumb to these things?) is dubious. I'm told that herpetologist Carl Gans, a faculty member and former chairman of the Biology Department of the University of Michigan, was so honored, in a spoof, when he was a precocious graduate student at Harvard. To be sure, I was pleased when Jim Mead and Larry Agenbroad broached plans for a gathering at Mammoth Hot Springs, South Dakota, one of my favorite places. Then they invited me to write something, "... if you want," said Jim Mead, in his offhand way. I took this as a chance to shed some light on the Desert Laboratory, my home port, and the spawning ground of most of the guests at the Hot Springs gathering, as well as many that were not.

That's the rub. The location for our rendezvous meant that only a sample of the many alumni and friends of "The Hill" could be on hand. "The Hill" is Tumamoc, rising to almost 3000 feet above St. Mary's Hospital and overlooking the west side of Tucson in southern Arizona. "Tumamoc" is Tohono O'odham (Desert Papago) for horned toad. Founded in 1903 by the Carnegie Institution of Washington as a desert botanical laboratory, the buildings became the launching pad in the late 1950s of the Program in Geochronology at the University of Arizona. The grounds comprise 870 acres (350 hectares) inside an ancient fence which is kindly honored by our neighbors. The vegetation is Sonoran Desert with creosote bush, palo verde, saguaro, brittle bush and 300 other species of native vascular plants along with 50 alien species (Burgess *et al.*,1991). There are permanent plots, repeat photo stations, maps of individual saguaros going back to 1909, and other data of unique value to desert ecologists. Currently, the Hill is being rejuvenated as a "Desert Laboratory" by various departments in the University and continues to be an academic ashram for those fortunate to be in residence. Wildlife ecologist Dave Brown calls the Desert Laboratory the "*Jerusalem of Desert Rats*." Author Larry Clark Powell says, "If Tucson has an Acropolis, this is it, the "Hill of the Horned Toad."

I hope my sketches of some of the post-Carnegie research ferment on and off the "Hill" will entertain not only the principals but also friends and supporters. Even members of the Mammoth Site who do not know the Desert

Late Quaternary Environments and Deep History: A Tribute to Paul S. Martin, edited by David W. Steadman and Jim I. Mead. The Mammoth Site of Hot Springs, South Dakota, Inc. Scientific Papers, Volume 3. Hot Springs, South Dakota, 1995.

Lab on Tumamoc may take an interest in the place where Larry Agenbroad and Jim Mead cut their teeth. Finally, those devotees of the Desert Lab who find I have overlooked the really important events in the last few decades on the Hill or, worst of all, failed to mention their names, will have to settle for a heartfelt *mea culpa*. I am forgetful, and I often treat my favorites with less care than strangers. The Desert Lab deserves more than one voice. I encourage you to *schrift* for yourself.

The geochronology decade: 1956-1965

Organized research on the environment of the arid west began on Tumamoc Hill when Carnegie botanists undertook their studies over ninety years ago. Their findings yielded a rich harvest on physiology, taxonomy, distribution, community composition, and demography of desert plants (McGinnies, 1981; Bowers, 1988). What transpired after Carnegie and the U.S. Department of Agriculture (USDA) interregnum is poorly known beyond our portals. Outsiders may not be fully aware of the research on historical ecology ("deep history") and related fields based on new approaches to the fossil record of arid regions and emanating from Tumamoc and its Desert Lab in the second half of this century. It was the contribution of an energetic and enthusiastic cadre of University of Arizona graduate students, research associates, sabbatical leave visitors, and some faculty, joined by a small and very productive U.S. Geological Survey team. The devotees, the graduate students especially, loved the place. Their research resurrected a reputation established by our predecessors.

I first visited Tumamoc Hill in November 1957 when my new boss and the driving force behind the University of Arizona's new program in Geochronology, Terah (Ted) Smiley, invited me to his office which we would share. It is now the library and conference room in the main building. We discussed details of setting up our new National Science Foundation (NSF) grant on ice age and postglacial pollen analysis. Our budget would cover salaries (mine was $6100 a year), supplies and radiocarbon dates, a crucial ingredient in our proposal. I was fresh from the east coast, very glad to have a research opportunity in Arizona and very happy to be close to México where I had done some fieldwork and expected to do more.

Then, as now, "The Hill" featured a collection of three small stone buildings with three younger metal "tin" buildings immediately to the south. The Geochronology Program shared space with S. Clark Martin and Herb Hull of the USDA, both highly able botanists and researchers that we were glad to have as neighbors. The grounds were secured for the University in the late 1950s by legislation introduced by then Arizona Senator Carl Hayden on the initiative of then University of Arizona President Richard Harvill. Later, "The Hill" had two narrow escapes from assaults on its integrity. The first was the proposal to level the old stone buildings and replace the Desert Lab with a teaching hospital. Phil Krutzsch, a Professor of Anatomy, was on the Site Committee and successfully opposed the idea. The second was the proposal to move married student housing at "Polo Village," the grounds the University's teaching hospital would come to occupy, out to the Desert Lab. The Nature Conservancy helped scuttle that scheme.

At the lab I found Ted puffing his pipe behind his desk ready with a big surprise. My mind was on this beautiful old building enmeshed in desert. Ted handed over the contents of some paper bags filled with dry material that looked suspiciously like some sort of manure ("road apples"). The road apples looked too big to attribute to a horse.

"Have you ever seen anything like this?" Ted inquired. I certainly had not. "It's ground sloth coprolite," Ted explained. (In those days we didn't say "shit.") The fossil manure had been carefully collected in stratigraphic sequence by anthropologist Richard Shutler from a dry cave in the western Grand Canyon. The samples had a distinctive pungent odor that was not offensive. One bag held pack rat midden debris including lozenge-size pellets of an artiodactyl which we later decided must be dung of Harrington's extinct mountain goat. The midden also yielded cactus aureoles and twigs of a conifer that UA taxonomist Charles T. Mason identified as Cupressaceae (juniper or Arizona cypress). Unwittingly we were looking at the dung deposits of extinct animals along with plant fossils of the last ice age. Much future research would follow study of similar material. The old Desert Lab seemed to be spinning through a time warp. I was enchanted. With such fossil material, new light could be shed on late Pleistocene climates and the mystery of late Pleistocene extinctions.

To initiate pre-Pleistocene palynology, paleobotanist Jane Gray arrived on the Hill on a Rockefeller Foundation grant to the University. She helped me cobble together a pollen diagram from the dung samples. Dick Shutler generated radiocarbon dates from Lamont Geophysical Laboratory and Bruno Sabels, assisted by geochemist Paul Damon, contributed chemical analysis of trace elements. Our findings did not reveal any obvious dietary explanation for sloth extinction, including trace metal deficiencies, one of the notions at the time.

Much happened in the next twelve years including travel to Poland, Costa Rica, Africa and Madagascar and I did not actually enter Rampart Cave itself until 1969. I flaunted a cardinal rule for the paleontologist: "Never publish on an important fossil locality that you have not seen yourself." Later I discovered how little time there would be to do research in Rampart Cave.

I did manage to make a field trip with Bernie Arms, our lab assistant, to Bat Cave below Quartermaster View just upstream from Rampart Cave. To mine the guano, the U.S. Guano Corporation built a million dollar tramway from the Hualapai Reservation to the cave across the Colorado River, packaging the guano for retail sale out of surplus buildings at the Kingman airport. We had learned of the project from an article in *Life Magazine* and from ads in local newspapers.

Bernie and I thought we might find interesting fossils, perhaps even more ground sloth dung in Bat Cave. We contacted Mr. Varley Compton of the U.S. Guano Corporation and were promised a chance to inspect their operation. In their cable car we swung off the rim of Quartermaster View and descended half a mile into the Canyon to find ourselves dangling above the Colorado River. From the far side landing tower a much smaller cable delivered the miners to the cavernous mouth of Bat Cave. Inside, all the deposits we saw looked like bat guano. Fresh guano reeks of ammonia and it is unlikely that large mammals such as ground sloths would inhabit the premises when guano was raining down.

The only fossil bones we saw were small, evidently those of *Tadarida*, the Mexican free-tailed bat of Carlsbad Caverns fame. I was disappointed when the guano yielded no fossil pollen, only wing scales of small moths and exoskeletal parts of beetles of no paleoclimatic significance. One guano sample was around 12,000 radiocarbon years old, old enough to be contemporary with the Rampart Cave ground sloths. We were in the right region and sampling the right time frame. Our breathtaking ride into the Grand Canyon by cable car entailed a means of transport that inflicted minimal damage on the land while providing ideal interpretive opportunities. Soon after our visit, an Air Force jet from Nevada buzzed the Inner Gorge and struck the cable with its tail. Somehow the jet survived, although the cable did not, and the system was abandoned. Tower parts are still in evidence. As far as I know, no other paleoecologists visited Bat Cave during the mining operation.

To showcase our program Ted started a Geochronology Colloquium, with tea, on Friday afternoons. A small and enthusiastic interdisciplinary mix of geologists, biologists and archaeologists could be expected. One day Ted told me that "Dr. D" wanted to attend our next colloquium and would I drive him out from the Tree-Ring Laboratory on the campus? I would. At 89, A.E. Douglass was a bit unsteady but quite alert and when we reached our gate at the foot of the Hill he wanted to make sure no traffic was coming down! Dr. D's last previous trip to Tumamoc, I learned, predated the construction of the present paved road, which is narrow enough to be sure, but not one-way like the one he remembered from the 1930s.

The Hill hummed with activity. In the pollen lab, totally rebuilt with NSF funds, Bernie Arms extracted alluvium from archaeological sites and a botany grad student, Jim Schoenwetter, helped count the pollen. At Matty Canyon in the Empire Valley southeast of Tucson, anthropologist Frank Eddy and geologist Spade Cooley were studying a beautifully stratified deposit. Over 5 meters of vertical exposure with buried organic lenses of ancient cienega deposits looked ideal for radiocarbon dating of our pollen profiles. The profiles accompanied archaeological remains, including pit houses.

Al Traverse, a Harvard-trained paleo-botanist then employed by the Houston Development Lab of Shell Oil Corporation offered us some free radiocarbon dates. With his classmate, Jane Gray, located on Tumamoc, Al decided to come for a visit. If not a rival of the Grand Canyon, Al assumed Matty Canyon must at least be scenically impressive or it would not be called a canyon. On our visit to the site Al discovered himself in the middle of a broad valley of mesquite grassland. He clambered down the banks of the arroyo, the entire section at most 25 feet deep and barely 3000 years old. Al kept his peace but I could see he thought Matty "Canyon" was a misnomer. He wondered if alluvial palynology was geologically old enough to take seriously.

While not old, the changes in the fossil pollen at Matty Canyon helped to initiate an alluvial chronology, one of our NSF research objectives. We were delighted to find some fossil pollen of maize (corn) in the older units, tangible evidence for prehistoric corn cultivation in Arizona. More maize (*Zea*) pollen appeared in a buried alluvial unit at Point of Pines where anthropologists Emil Haury and Ray Thompson ran the University of Arizona archaeological field school. Radiocarbon dates associated with the maize pollen ranged from 2300 to 4000 years old. With understandable pride, Jim Schoenwetter and I published our first paper in *Science* and caught some criticism from moss-backs for our breezy title: "Arizona's oldest cornfield." Years later, accelerator dates on corn cob fragments would solidly establish the age of *Zea* in the region. From the top of Tumamoc Hill itself, archaeologists Sue and Paul Fish have recently excavated bits of corn cobs of which the oldest was accelerator dated at around 2700 years ago. Bruce and Lisa Huckell also verified the idea that corn was being cultivated in southern Arizona before ceramics (pottery) came into use.

Archaeology aside, what I most wanted was a pollen record through sediments incorporating bones of extinct animals, ideally at a kill site such as the one at Lehner Ranch near Hereford on the San Pedro River. I wanted pollen evidence reflecting the environment experienced by the mammoths. For his Masters thesis archaeologist Lex Lindsay struggled with pollen extraction of samples from the Lehner Ranch, a mammoth kill site being excavated by the Arizona State Museum. I assured Lex that techniques I had learned in Ed Deevey's pollen lab at Yale incorporating hot concentrated hydrofluoric acid (HF) would do the trick. Admittedly, hot HF is not to be trifled with, and to handle it our centrifuge tubes had to be made of Nalgene, an acid- and a heat-resistant plastic. The method yielded countable pollen extractions from San Simon, an Arizona cienega that University of Arizona botanist Eddy Kurtz kindly helped me core in the summer of 1956. Surely, the new technique would help recover pollen from the beds yielding the intriguing fossil mammoth-Clovis point association at the Lehner Ranch. I was wrong. The extracted residues were milky white, with pollen grains obscured by a siliceous colloid that resisted further treatment.

Eventually Peter J. Mehringer, Jr. (Pete) managed to dissolve most inorganic colloids with concentrated nitric acid. Even then, the Lehner pollen samples were miserable to count, and the results (including scarcity of tree pollen) seem discordant with what I think Cochise County should have looked like over 10,000 years ago.

In the spring of 1962 Pete and C. Vance Haynes enrolled in my pollen course. As part of the learning experience they were assigned projector duty at the First International Pollen Conference which attracted some 200 scientists to the University of Arizona campus during Easter week break. The conference was the brain child of Ted Smiley, ever the visionary. I assured Pete and Vance that they would hear and learn about the most recent research being attempted in this new field. It turned out that the projectors of those days were too noisy to allow the operator to hear anyone not shouting in his ear!

Field trips before and after the conference were planned, announced and filled. At the time biometrician Jim Mosimann was helping us with statistical interpretations of pollen from the Willcox Playa. Bill Byers was a grad student, investigating fossil pollen from Glen Canyon and Wetherill Mesa in Utah. Jim, Bill and I teamed up with Roger Morrison of the U.S. Geological Survey to guide a group of twenty palynologists and botanists on a visit to the Cloud Forest at Rancho del Cielo in southern Tamaulipas, México, 1500 miles southeast of Tucson. We had gone to all the trouble because we felt that anyone studying fossil pollen had to see "anomalous" mixtures of trees with good early Tertiary fossil records, in this case temperate *Liquidambar* (sweet gum), beech (*Fagus*), beech drops (*Epifagus*), ironwood (*Carpinus*) and fir (*Abies*) growing with tropical *Podocarpus*, *Meliosma*, *Eugenia* and tank bromeliads in a moist forest whose terrestrial fauna curiously lacks the Appalachian affinity seen in the plants. The field party, which included ecologists Pierre and Francoise Dansereau, Estella Leopold, Loren Potter, Matsuo Tsukada, Bill Watts and various European palynologists, was properly smitten with the Rancho del Cielo Cloud Forest and its guardian, Frank Harrison. It took us three days of demented driving through the Chihuahuan Desert of northern México to get back to Tucson in time for the conference.

Glowing with the perceived success of the cloud forest excursion, I imagined the post-conference field trip to playa lakes and alluvial outcrops in Arizona, New Mexico and west Texas would be routine. Our first stop was at Whitewater Draw near Double Adobe, the type site of early Holocene alluvium in the region. Several 8000-year-old radiocarbon dates and fossil pollen had come from sediments that had once disgorged bones of mammoth and other extinct fauna. I would interpret the record to Doug Byers, Ed Cushing, Frederick Johnson, Jock McAndrews, Gurdip Singh, Herb Wright and other luminaries, some old hands at alluvial geology. I had read the literature and had no doubt that I could read the pollen record in the outcrop, having collected and counted it myself. Pete Mehringer and Vance Haynes dutifully accepted the task of cleaning up the sections at Double Adobe for our field inspection.

They also bore the brunt of the field interpretation when unexpected questions I could not answer were raised about the depositional gaps (breaks in the sequence) which I had not noticed. The idea began to grow on me that there might be problems with the pollen record at Double Adobe. For one thing, it was not securely associated with the extinct fauna. For another, it suffered gaps of unknown duration. In any case, it was not a kill site and eventually Vance worked out that, contrary to expectations, radiocarbon dates on mammoth sites with cultural material were all around 11,000 years old. Much later when Mike Waters trenched the Double Adobe flood plain with a back hoe I would better appreciate the complexity of the cut-fill deposit.

On the recommendation of archaeologist Ted Sayles, who had collected a mammoth tooth at the spot, Jim Schoenwetter and I collected a Holocene pollen section near Fairbank on the San Pedro River. Exploring the same wash, Pete and Vance discovered the Murray Springs mammoth-Clovis site. Sayles had what was probably a mammoth molar redeposited from Murray Springs. Mammoths at Murray Springs proved to be 11,000 years old, verifying Vance Haynes' chronology elsewhere. To our dismay, the section lacked fossil pollen. Aside from Lehner we were having trouble finding fossil pollen and fossil mammoth bones *in situ.* Many of the older alluvial outcrops proved sterile or virtually so. Jane Gray also was experiencing difficulties with fossil pollen in outcrops of older (late Tertiary) sediments. My enthusiasm for arid land alluvial palynology began to cool. My impression of the abilities of Haynes and Mehringer soared, especially after they applied their skills to the excavation of the Tule Springs site outside Las Vegas, Nevada, using new protocols for pretreatment of radiocarbon samples. As a result, the alleged pre-Clovis age of the site collapsed.

For fossil pollen in arid land sediments the black muds cored from the floors of "dry lakes" such as the Willcox Playa in southeastern Arizona were more promising. For one thing, the reduced muds were rich in well preserved fossil pollen and fossil algae. For another, the basins were deep. Hundreds or even thousands of feet of lacustrine sediments could be expected in the arid west, a land of block-faulted basins and ranges. With luck and a hefty budget for core drilling, it might be possible to build a pollen chronology through the entire Pleistocene, the last 1.7 million years. So far we had the upper 140 feet from the Willcox Playa, an undated drill core with well preserved pollen in many levels.

My lab assistant Harrison (Harry) Yocum counted much of the fossil pollen from the Willcox Playa. One quiet afternoon he sensed he was being watched. From a small hole in the ceiling Harry detected a nose, two bright eyes, then a head and ears of a ring-tailed cat (*Bassariscus*). The animal returned on successive afternoons and grew more bold, perching on top of a cabinet. When Harry fed it by hand, matters got out of hand. At night the ring-tail began to explore the lab. It jumped onto Jane Gray's drafting table, urinating on and ruining the final copy of a painstakingly drafted pollen diagram. Perhaps it thought the scent of India ink was the territorial mark of a rival; I don't know. Other outrages followed, including destruction of an expensive new pair of field boots and the slaughter in its cage of a pet parakeet named "Shreve." Finally a new policy was implemented. Ring-tails would not be fed and all access holes would be plugged.

Our pollen successes attracted talent. In the mid 1960s Dave Adam arrived on the Hill ready to explore long cores from western basins. We decided to try to recruit the help of a new addition to the University's Botany Department from Chicago, Willard Van Asdall. We heard that Will was interested in the Chenopodiaceae, including the salt bushes. They favor alkaline soils on the shores of dry lakes such as the Willcox Playa. They were major contributors to the fossil pollen record. We decided to introduce Will to the Willcox Playa. We asked my pilot friend Alexander

"Ike" Russell, who was taking me into the mountains of Sonora to collect plants around remote airports, to fly us to Willcox in his Cessna. When we got to the playa, roughly eight miles wide, Ike decided it was dry enough to land, first testing his opinion with a touch-and-go to make sure the surface was dry and hard enough to be safe. From our impromptu airstrip we could walk to meter-high mounds, wind-blown teardrops perched on the floor of the dry lake. The mounds are formed of clay pellets accumulating in the lee of bushes of the shrubby chenopod *Suadea*. While Will was suitably impressed, he found better things to do than tackle the ecology of dry lake salt bushes. In coming years, through Will's guidance, a sizable and very talented group of ethnobiologists graduated from the University.

In 1962 Jane Gray left for a tenure-track position at the University of Oregon. She was succeeded on the Hill by G.O.W. (Gerhard) Kremp, formerly with the U.S. Geological Survey. One of his students, Al Gottesfeld, finished a master's thesis on the terrestrial paleoecology of the Triassic at Petrified Forest. Al loved the desert and camped out in it whenever he could. One night he called to tell me to come up to the lab with an ice chest. He had returned from Rocky Point on the Gulf. After witnessing a fabulous grunion run at the height of spring tide, Al had shoveled grunion and ice into the back of his truck and raced back to the Hill to share the fresh catch with all comers.

While the Geochronology experiment would soon be folded back into the Geology Department, I viewed the first years as very promising. We had established the potential of and some limitations of Quaternary palynology in arid regions and uncovered new evidence for glacial age cooling at low latitudes. Dave Adam and Jim Mosimann made pioneering contributions to the statistical treatment of pollen counts. To traditional arguments that the extinction of late Pleistocene megafauna was driven by climatic change, Pete and I had probed an interesting alternative, "overkill." Graduate students from various parts of the country found their way to the Hill, such as ethnobotanist Vorsila Bohrer, geologist Charles Schweger and entomologist John Mathews. Charlie and John were both interested in Beringian deep history. An Argentinean paleobotanist, Carlos Menendez, and a Mexican palynologist, Lauro Gonzalez, came on sabbatical leave visits. I was now eligible for one of those myself.

Sabbatical I. East Africa and Madagascar

Supported by a Guggenheim Fellowship, my first sabbatical which began in 1965 brought the opportunity to gain a wider view of both the Desert Lab, of arid land ecology, and of megafaunal extinction. I wanted to see the "living megafauna" in its African heartland and to visit some fossil sites, especially Olduvai Gorge being dug by Louis Leakey. My close friends, Ike and Jean Russell, decided to come along in their Cessna. We could visit more localities by air than on the ground. They would help with field and lab work. I was overjoyed.

Before we left I would participate in the International Quaternary Association Congress of ice age geologists from around the world gathering in Boulder, Colorado. From a symposium on "Pleistocene Extinctions" Herb Wright helped me edit a book which featured chapters by Ed Deevey, John Guilday, Vance Haynes, Art Jelinek, Estella Leopold, Ernie Lundelius, Pete Mehringer, Nikolai Vereshchagin and other friends of the Desert Laboratory. Led by Leo Heindl, there was a post-Congress field trip by bus through various parts of the Colorado Plateau. Most of the stops interpreted surficial geology. While scenic, many of the experts wanted more. They wanted to see a fresh exposure of an important site. At Lehner Ranch the conferees were delighted to see trenches exposing dated units and fossil pollen embracing the famous fossil mammoth-Clovis association, all carefully prepared by Pete Mehringer and Vance Haynes. Ed and Lynn Lehner, the genial and talented owners of the site, hosted the crowd and added the names of a variety of eminent out-of-state and foreign archaeologists and geologists to their guest book. The years would see many more such luminaries coming on pilgrimage to the Lehner Ranch on the Upper San Pedro Valley. Most tourists to the region are interested in Tombstone and the shootout at the O.K. Corral. The spear-out of mammoths by the first Americans has yet to find its public.

In October I left my family in Tucson and flew to Cairo via England. In Cambridge I gave a seminar on my favorite subject, Pleistocene extinction, to Harry Godwin's Subdepartment of Quaternary Research. On arrival in Cairo my BOAC jet flight was delayed a few minutes by traffic. The traffic, it turned out, was Ike's Cessna cleared to land ahead of us at the Cairo airport. Ike and Jean had shipped the Cessna by boat across the Atlantic and flown it from England through Spain to Egypt. For the flight to Nairobi we were joined by a friend of the Russells, Major William Swan, U.S. Army (Ret.). With the four of us (and lots of excess baggage), the Cessna was truly overloaded.

Departing from our airway instrument flight plan to Aswan, we flew "by sight" up the Nile. Landing at twilight at Luxor we unloaded, tied down, and then, to the astonishment of airport guards, proceeded to unroll our sleeping bags and to set up camp under the wing of the Cessna in a corner of the field, as was the pilot's custom. There were military planes around and this wasn't a lonely strip in the Sierra Madre of Chihuahua. "Don't you intend to spend the night in one of the tourist hotels in town?" We did not. The airport guards finally offered a compromise. We could sleep out at the base of the tower next to a barracks under mercury vapor lights, which we reluctantly did. The mosquitoes were very bad and so was the idea of camping out at Luxor Airport.

At Aswan Ike discovered he had burned out his left wheel brake. There was no chance for repairs. We left for the Sudan and flew across Nubia, a vast desert wilderness worthy of the name with no trees or shrubs or ground cover of any type except along the Nile. This was the hyperarid region, as different from the Sonoran "Desert" as Antarctica is from Patagonian steppe. The Carnegie desert lab scientists Godfrey Sykes and D. T. McDougal had done research here.

On take off at Malakal two maintenance workers cutting grass suddenly loomed up unexpectedly out of a ditch next to us as we roared down the strip. Fearing that he might hit them, Ike jammed on the brakes and cut power. While the bad brake did not work, the good one did, all to well, throwing us into a ground loop which crumpled the wing tip and tail tip and collapsed a wheel. In a week Ike somehow managed to repair the damage well enough to ferry the Cessna on to Nairobi with Jean while I flew ahead on Sudan Airways and Major Swan returned to New York.

The Cessna spent three months in the shop at Nairobi undergoing repairs while we visited game parks and attempted some fossil pollen extractions at William Isaac's botany laboratory in the University College of Nairobi. Prof. Isaac and his wife were genial marine algologists, and parents of the famous archaeologist Glen Isaac. Bill Bishop of the Kenya Geological Survey directed us to outcrops around Lake Magadi which yielded vast numbers of fish bones trapped by an ash fall along with some fossil pollen.

Africa had much to offer. Now the television is awash with nature shows on African big game. In the 1960s, it was all much more remote. For reference material we collected megafaunal manures. We sampled thorns on fever trees, detecting reduction in thorn number with height. We visited Lake Turkana,

Serengeti Park and near Uaso Nyiro we met George and Joy Adamson.

Louis Leakey invited me to ride with him on trips to Olduvai Gorge from Nairobi. I tried and failed to persuade him that early Pleistocene extinctions of carnivores, giant pigs and giant baboons might be triggered by competition with or even predation by early hominids. I was impressed with Leakey's success at exhibiting early man-extinct megafaunal sites at both Olduvai, Olorgesaille, and Kariandusi. Although the Hot Springs, South Dakota, display is superior to anything of its kind in commemorating America's extinct mammoths, there still are no Clovis-mammoth associations on public exhibit in the United States to match what Kenya offers. Our Eurocentric approach to New World patrimony glosses over the significance of Clovis colonization. Where is the heritage memorial to honor the first Americans themselves?

Topping off the trip was a month spent in southwestern Madagascar. The Cessna was now repaired and ready to go. We met anatomist Alan Walker in Kampala, Uganda, persuaded him to join us, and flew for two hours through solid overcast east from Nampula across the Mozambique Channel to Majunga, where much to our relief the clouds parted. We could see red flood waters coloring the ocean and then the Madagascar coast itself. Near Tulear in southwestern Madagascar, we visited beaches littered with shells, not of clams but broken egg shells of the elephant bird, A*epyornis*. We visited fossil deposits in Holocene alluvium, similar to arroyo banks to be found in Cochise County and throughout the west. The difference was that in Madagascar the deposits might harbor bones of extinct animals, including hippo and giant tortoise. The extinct fauna disappeared very late, only a thousand years ago or so, coincidental with human colonization, although butchering sites of the elephant bird and other extinct animals are curiously scarce.

The Africa-Madagascar experience was worth a follow-up. I returned with my family in the summer of 1966 to revisit Tsavo and to spend another month in Madagascar with Tucson botanists Richard Felger and Pierre Fischer at the famous French Government Marine Field Station at Tulear. We collected fossils of extinct hippo for the Smithsonian Museum and more egg shells of *Aepyornis* which turned out to be ideal for radiocarbon dating.

Fossil rat middens: "Gold in them thar hills"

Back at Tumamoc Pete Mehringer was teaching the pollen course and had joined the Geochronology faculty. We both were impressed with Phil Wells and Clive Jorgensen's success at chronology building in the Mojave Desert using fossil rat middens. We knew Rampart Cave harbored similar material. What about other parts of Arizona and the Southwest? In a new course "Paleoecology and Man" I found an opportunity to coax graduate students shopping for a thesis to join the treasure hunt, the search for ancient middens. The glacial age pollen record which Dave Adam, Anne Bent, Kathryn Clisby, Ulf Hafsten, Dick Hevly, Estella Leopold, Pete Mehringer, Paul Sears, Herb Wright and others had pioneered in this region suggested important change in glacial times. Desert grassland appeared to have been overrun by pine forest during the full glacial, if we read the fossil record right. But no one knew what went on in the low elevation deserts, the sort of record that the Carnegie botanists would have wanted. Wells and Jorgensen said there had been a juniper invasion in California's deserts and we had that twig of juniper from Rampart Cave in Arizona's segment of the Mohave Desert. What happened to the saguaro?

Geochemist Paul Damon tried his hand at prospecting. To our astonishment he brought in remains of a fossil midden with juniper (and lacking saguaro) that he had collected from a small rock shelter near Picture Rocks Pass in the Tucson Mountains, not far from the Desert Laboratory. Saguaro, not juniper, grows in those mountains today. This was the imprint of the ice age changes we had been seeking, and here it was right in our back yard!

The Great Fossil Rat Midden Hunt now began. A "Corps of Discovery" spearheaded by Tom Van Devender and including Martha Ames, Jim King, Paul Leskinen, Pete Mehringer, Kevin Moodie, Art Phillips, Geof Spaulding, and Jeff Zauderer were joined sooner or later by Julio Betancourt, Ken Cole, Pat Fall, Lisa Graumlich, Cynthia Lindquist, Jim Mead, Bob Thompson, Larry Toolin, and Bob Webb, among others. While the Tucson Mountains were not a rich source, lots of fossil midden sites were found in western Arizona, eastern California, Nevada, southern Utah, New Mexico, west Texas and northern México. It became apparent that in addition to juniper the southwest deserts had been invaded by scrub oak, pinyon, sagebrush, yucca and shadscale. At higher elevations there was another shuffle, verifying what we found in the fossil pollen record. Montane conifers such as limber pine, Douglas fir, white fir, dwarf juniper, and occasionally spruce or bristlecone pine occupied today's woodlands of pinyon-juniper or oak-juniper. Clearly, the biota of the arid west had been profoundly affected by climatic conditions over the last 40,000 years and in their papers in *Science*, *Nature* and other professional journals the grad students at the Desert Laboratory led the way to new finds in paleoecology. Whenever a student had a worthy manuscript on fossil middens and needed some extra financial help to see it into publication, support could usually be found in an administrative office.

I think it was December of 1968 when Robert MacArthur, perhaps the best of the new breed of American ecologists of the time, came to Tucson to spend part of his sabbatical at Tumamoc Hill. Pleistocene extinctions had begun to attract media attention, including a piece in the December 7th issue of *Time*. Robert helped contribute to an advanced ecology seminar on extinctions which I ran jointly with faculty from the Biology Department.

Around then Dave Adam left us for the U.S. Geological Survey in Menlo Park, a great research opportunity for him given Dave's interest in the fossil records of drill cores from western lakes and playas. Pete Mehringer left for the University of Utah and eventually became Professor of Anthropology at Washington State University in Pullman.

Pete's final contribution before departing was the Missouri mastodon project. Not since the time of A.C. Koch (in the 1840s) had there been such excitement about mastodons in Missouri, an unlikely source for such fossils. They are more often found in Michigan and Florida. It all began when serendipity lent a hand.

On a summer contract Pete had tried and failed to recover fossil pollen from Rogers Shelter in the flood zone of the soon-to-be-constructed Harry S. Truman Reservoir. Seeking something to show for his bid, Pete turned to an adjacent spring-fed marsh on the terrace of the Pomme de Terre River, a place called Boney Spring. The cores he recovered looked very promising for fossil pollen. Then Pete noticed they contained bone and tusk fragments. "That's from my cow. She bogged down here years ago," commented the skeptical owner of the spring, a Missouri dairy farmer who was watching the coring operation.

It was not a cow at all; it was bone from not one but what eventually turned out to be 31 individual mastodons in a bone bed of 731 bones between 13,600 and 16,200 years old. It was grist for two dissertations, one by Jim King and the other by Jeff Saunders, and field experience for many University of Arizona students. Among other finds the Missouri mastodon team laid to rest A.C. Koch's 150-year-old claim that Missouri mastodons had been hunted by prehistoric people. The mastodon bones were much older than the artifacts, although some artifacts were sufficiently close to the fossil bones that it was easy to see how they might mislead the unwary. Unfortunately, the fossil mastodon bone bed could not be permanently exhibited *in situ*.

By 1969 a new palynologist, Al Solomon, had arrived on Tumamoc to initiate modern (airborne) pollen sampling and to join the new scanning electron microscope (SEM) group. Our SEM opportunity came when a friend brought Charlie Drew to Tumamoc Hill. Charlie was a chemist working as a research scientist at the Navy's Michelson Laboratory located at China Lake, California. Charlie had recently acquired one of the first scanning electron microscopes, a vast improvement over the light microscope in examining surfaces. He had heard that pollen was an ideal subject for SEM study and he was right. SEM photomicrographs were stunning and soon appeared on covers of all sorts of scientific journals. With the help of Al Solomon and Jim King, Charlie and I published an atlas of southwestern pollen types. Promising as it was, the method was too cumbersome, at least in our hands, to apply to routine fossil analysis and hoped-for breakthroughs in paleoecology were not forthcoming. Much better use of the SEM was made by Donna Howell who photomicrographed the brush-like tongue of a long-nosed bat, showing how it was able to mop up pollen from its face and head as well as from its favorite flowers, those of agave and columnar cacti. When she was studying their diet, Donna briefly kept an experimental bat colony in the attic above the main lab. If Tumamoc lacked a belfry, at least for a short while we had the bats.

In 1969, guided by Doug Evans of the National Park Service, Austin Long and I finally got around to investigating the ground sloth caves in the Grand Canyon, beginning with Rampart and Muav. That summer we also accepted Bob Euler's invitation to join in the excavation of Stanton's Cave. There was ample material for research on fossil pack rat middens, diet of extinct animals, and the chronology of megafaunal extinction. 1969 also saw our first Grand Canyon Colorado River trip sponsored by the Arizona Academy of Science. Near Shinumo, Jim Mead, then a high school student, discovered fossil pack rat middens (with displaced juniper) in the Inner Gorge, the first fossil juniper from the Canyon since Dick Shutler's collection from Rampart Cave. Midden collections from Stanton's Cave yielded ample juniper, twig figurines and occasional fossil birds including vulture bones. Our river trips anticipated things to come. Lounging comfortably on rubber rafts while floating downstream, we gazed up at inaccessible holes far above us in the Redwall Limestone, idly speculating that some should have been the nesting sites of the extinct California condor. It turned out that we were right, as Steve Emslie and Jim Mead discovered later after investigating some of the "inaccessible" caves.

Grand Canyon research appealed to Jim King, Jim Mead, Art Phillips, Wayne Sigleo, Geof Spaulding and Tom Van Devender. In an extraordinary bout of field work in the roughest and most physically demanding parts of the Grand Canyon, Ken Cole took advantage of his skills as a gymnast to recover suites of midden samples that revealed significant change in all the common tree and shrub species from the Canyon rim to the sandstone cliffs just above the Colorado River. Someone told Ken his new data was good enough to publish in *Science*; he submitted a paper and *Science* published it. Ken's discoveries included dung and bone deposits of the extinct goat, *Oreamnos harringtoni*, material that helped Jim Mead with his dissertation on that animal.

In the Sheep Range of southern Nevada Geof Spaulding elegantly displayed the shuffle of montane conifers moving to higher elevations and being joined by extralocal species in the Holocene, a biotic enrichment as the climate warmed. In the Great Basin Bob Thompson combined fossil pollen with midden analysis to show that while the vegetation of the last ice age supported arid-adapted limber and bristlecone pines down to the elevation of Lake Lahonton, the trees were scattered among shrubs and probably did not constitute a forest. Even the last 1000 years showed slight but significant translocations from black brush to creosote bush in the Mojave Desert where Bob Webb investigated desert plant recovery on abandoned and cleared sites of 100-year-old ghost towns. Julio Betancourt found major changes in vegetation of the Four Corners area. Last but not least, Tom Van Devender ransacked desert mountains from Coahuila and west Texas to eastern California and Sonora discovering a rich late Quaternary record of previously unknown woodlands occupying what are now treeless desert mountains. Tom noted a number of distributional overlaps, species of southwestern plants coexisting in middens and presumably in nature that no longer associate as far as botanists are aware. Some truly incredible associations were put to the test by AMS dating and found to be the result of sample contamination. Still, the record of anomalies seems valid and was supported by direct dating in at least a few cases, as Tom and the "Corps of Discovery" showed in a paper in *Nature*.

In 1972 Dick Harington, an expert on *Oreamnos*, came down from Ottawa to see the fresh-looking dry pellets and other fossils of extinct goats in place at Rampart Cave. Two globetrotting Australians, palynologist Geof Hope and paleontologist Jeanette Hope, inspected Rampart on one of our trips; later Geof returned to Tumamoc on a sabbatical. Both would help me visit key fossil sites in Australia. Fran Bartos undertook pollen analysis of deer pellets and deer diet on Tumamoc Hill. Eleanora (Norrie) Iberall finished a thesis on diet of Harrington's extinct mountain goat based on the wealth of fecal pellets that Bob Euler had excavated from Stanton's Cave. In her literature search she discovered the materials analysis lab of Dick Hansen of Colorado State University. Dick visited the Hill, taught us more about how to "know your shit," and completed a study concluding that globe mallow, Mormon tea and salt bush were the main forage of extinct ground sloths. The sloth dung was so well preserved that it matched the carbon and nitrogen ratios of modern cow manure.

Gerald Kelso took fecal palynology to its nadir with a dissertation on the distribution of ingested pollen in human excreta, both fossil and fresh. Some found reason to protest the outrage when Gerald dried his odoriferous research samples in the lab's stove, previously devoted to warming casseroles at potlucks.

Austin Long, Dick Hansen and I published our new data from the Grand Canyon ground sloth caves, including radiocarbon dates that faded away around 11,000 years ago, coincidentally, we noted, with the age of the Clovis culture and extinct mammoth. Caves with ground sloth dung were also known in South America. When did the South American ground sloths vanish? More to the point, when did the last dung drop on the floor of those caves? That question was enough for Austin and me to begin planning field work in Chile and Argentina.

Sabbatical II. South America

In the fall of 1972, I was once again eligible for a leave. I had recovered from the terrors of the over-water flight to Madagascar and the ground loop in Malakal and proposed another grand tour of fossil sites of another continent, this time South America. Ike and Jean had flown from Tucson to Tierra del Fuego in their Cessna once before and they were ready to go again. Along the way, we narrowly escaped the Managua, Nicaragua, earthquake of October 1972 after just making it into Tapachula, on the México-Guatemala border, having glided into the field dead stick with empty gas tanks. Ike disliked the trouble of gassing up unnecessarily.

In Perú we visited Pichimachai Cave excavated by MacNeish and the famous Talara tar pits, with extinct animal bones exposed in fresh diggings, considerably more impressive than what one can see in California at Rancho La Brea. In Brazil we examined mastodon fossils at the Araxa Hot Springs, and in northern Venezuela some mastodon sites in spring deposits allegedly with cultural material. Nevertheless, the main object was the inspection of two caves known to harbor ground sloth dung. One was the famous Mylodon or Eberhardt's Cave in Chile, the other the recently discovered Gruta del Indio in Argentina, being investigated by Humberto (Tito) LaGiglia of San Rafael.

San Rafael is a beautiful small city in the wine district of Mendoza province. The surrounding desert looks very much like the eastern Mohave of California and Nevada. There are various desert shrubs that made us feel at home, especially creosote bush which unlike creosote bush in North America comes in more than one species.

Hector d'Antoni, a palynologist from Museo de la Plata, brought his family to San Rafael and joined us in the field. He was working on a pollen diagram from the dung in Gruta del Indio. Unfortunately, we could never be sure what type of ground sloth was involved. There were many extinct ground sloths in the fossil record of South America. Teeth of ground sloths of two different families were found in the cave. The "bonegas" (lumps of dung) were smaller than any Austin and I had seen in the U.S. and much smaller than the specimens collected at the Mylodon Cave in Chile. Their source remains a mystery. Whatever the species, the last dung we could find to date was around 10,000 years old.

Ike, Jean and I persuaded Tito to fly with us to Puerto Natales in Chile to collect samples in the Mylodon Cave, a huge cavern with a sizable deposit of manure looking out on glaciers of the Southern Andes across Ultima Esperanza Sound. The cave was made famous around 100 years ago. The hide and fresh-looking dung of the extinct mylodon triggered a futile search for living ground sloths. Recently cryptozoologist David Orim has resumed the search based on reports of a small ground sloth in the Upper Amazon. I wish him well. I cannot find evidence of ground sloths after 8000 BC. I think Orim started his search ten thousand years too late.

On one of our flights from the Natales airstrip to Punta Arenas, the overcast turned squally and Ike decided it might be wise to land. Besides, we were hungry and it was past lunch time. He spotted a sheep ranch at the end of a narrow dirt road straight enough to serve as an emergency strip. The wind was right and we tried it. Unfortunately, none of us had seen a line of small homemade telephone poles which converged on the road just ahead of us. Braking hard, Ike still could not stop the Cessna in time. Fortunately, the wood in the pole we struck was rotted and on impact the pole flew into pieces, leaving no more than a cup-shaped dent in the leading edge of our wing. By the time all had been explained to a troop of curious sheepherders from the station, we were truly hungry and accepted their hospitality, a steaming bowl of hearty mutton soup and mounds of tasty fresh baked bread. That afternoon the weather improved and after negotiating cost of damages to the phone line we flew on our way.

From Tierra del Fuego and its flocks of guanacos we flew toward the ice fields of the Darwin Channel until the weather turned foul, then north up the east coast. At La Plata Hector and Suzanna d'Antoni showed us Ameghino's world-famous circular museum. After helping collect a sample of Mylodon hide for ^{14}C dating, Tito LaGiglia returned to San Rafael. In Brazil Jean insisted we take time off to experience carnival in Rio where we were joined by my wife, Marian, who flew down from Tucson. On walking through narrow streets packed with wildly celebrating revelers en route to the main pavilion, Ike had his pocket picked. He saw it happen and could do nothing. In the mad carnival crush his arms were pinned to his sides.

Back at the Desert Lab in Tucson, Al Solomon had invited the genial Rutgers ecologist Murray Buell, recently retired, to teach a semester of my field ecology course. Murray and his wife loved students and hosted ice cream parties in their home. The parties became potlucks and, with Al Solomon's guidance, an eclectic self-sustaining student-run seminar, dubbed the "Mescal, Marching and Menudo Society" was born. Beer was consumed much more often than mescal, once or twice menudo (soup of tripe) was actually served, and plenty of marching back and forth was the fate of seminar speakers who were often obliged to run a gauntlet of sharp questions, rather like a cross examination in court. Many grad students would discover that their dissertation defense was a simple matter compared to their trial by fire as a guest speaker of the Menudo Society.

One visit to Argentina deserved another. In 1974, I went back with Tom Van Devender and my son Tom Martin to search (without success) for more ground sloth caves in Mendoza Province. Tito LaGiglia did show us organic deposits that looked like pack rat middens or the South American equivalent. Austin Long came down on his sabbatical and clarified the dating relationship between sloth extinction and earliest cultural activity (in the form of charcoal) at Gruta del Indio. In Argentina, as in North America, the overlap between first hunters and last ground sloths was very brief. Fossil middens and fossil pollen records showed change, not necessarily for the worse, during the time of extinction. The ground sloths must have been highly vulnerable to human predation whenever humans were on the scene. The year of 1974 was memorable, although I did not realize it at the time, for the discovery in Hot Springs, South Dakota, of the first bones of the Mammoth Site and the beginning of a unique, precious collaboration between the town of Hot Springs and a group of geologists led by Larry Agenbroad of Northern Arizona University.

Spirit of 76

In 1976, in response to archaeologists who seemed to believe nothing important could happen rapidly, Jim Mosimann and I modeled a maximum rate of impact by Clovis foragers which we called "blitzkrieg." It blitzed through the media with great success, topped off by a *Reader's Digest* article on our Rampart Cave research by James Michener. Nevertheless, many archaeologists rejected blitzkrieg. They expected to find people in the New World well before Clovis time 11,000 years ago and refused to accept the possibility that all megafaunal extinctions occurred together, as our model required. They certainly did not buy the idea that extinctions could happen in only a few hundred years. I was amused at the subsequent popularity of the asteroid (bolide) model for dinosaur extinction which doomed large animals in a much shorter interval of time, perhaps less than a year. Well, we had our human bolide timed with mammoth extinction: one good catastrophe deserved another.

On a hot July evening, as so many are in Tucson, the "Menudo Society" was enjoying a seminar around the swimming pool at Betty and Bill Fink's comfortable condominium. Betty was our superbly talented secretary and coordinator on the Hill. The potluck dishes, many vegetarian, were excellent. The speaker, Donna J. Howell, had an impressive command of her subject, the adaptations of pollinating bats, and her audience was unusually quiet and attentive. I was not happy to be called away to the telephone.

It was Mary Kay O'Rourke with an urgent message. My friends, Bob Euler and R. Roy Johnson, both researchers at Grand Canyon National Park, were trying to reach me. Then the blow fell, something about a report of smoke coming out of the mouth of Rampart Cave.

The slow combustion, hardly a "fire," lasted months, enduring all best efforts of Park Service crews and other firefighters to extinguish it. The Bullhead City Department volunteered to a man. The Tucson Fire Department conducted experiments on how to smother slow combustion. I learned about brattice cloth, employed to seal off the air. At one point, Park Service fire fighters in gas masks working inside the cave almost had it out. Then a rockfall threatened their safety and underground efforts ended. An exhaust system was attempted, pumping exhaust gas into the cave. Although the rate of combustion could be greatly reduced by oxygen depletion, the heat of combustion was retained in the dung blanket. When eventually fresh air leaked back into the cave, as eventually it did, combustion resumed.

All agreed the fire could be stopped by flooding. Since most of the fossils would be ruined in the process, I did not want that. Finally, six months after the fire was discovered, the Mine Safety group of the U.S. Bureau of Mines brought it under control. While a part of the deposit on the west (right) side of the cave had survived, the whole experience was a nightmare. In addition, I was dismayed (although not surprised) by snotty publicity. Here was a fossil deposit that revealed in magnificent detail the life history of several extinct animals otherwise only known by their bones, animals native to the Grand Canyon and to the Southwest to disappear forever not long ago, as geologists view the world. Nevertheless, some columnists could not resist the opportunity to trash the federal agencies for spending tax dollars in fighting a fire in "sloth dung." "What will they do next?" Well, I had livened up lectures with one-liners and now the columnists had their turn. I compared the loss of Rampart to the burning of the ancient library in Alexandria, with scant effect. During the Denver meetings of the American Association for the Advancement of Science that fall, when science was in the news, I was interviewed on CBS Television. CBS played the story straight for the most part, but could not resist an exit comment by Walter Cronkite, something about bringing their viewers the latest news on "endangered feces."

Not all was sackcloth and ashes. That year (1976) I learned that Ray Turner was looking for a new berth for his ecohydrology project in the U.S. Geological Survey. Ray had long demonstrated a genius at desert monitoring, including the interpretation of repeat photographs (Hastings and Turner 1965), the study of permanent photos established by Carnegie botanists and others on Tumamoc Hill (Goldberg and Turner 1986), saguaro demography, begun by Volney Spalding, and the detailed mapping of the range of Sonoran Desert trees and shrubs soon to be published. He more than anyone belonged on Tumamoc Hill. Now that Clark Martin was leaving the Hill and the Santa Rita Range Experiment Station along with range research was being terminated by the USDA, the ghosts of the Carnegie botanists would have given me no peace if Ray had not been urged to occupy space now available. Eventually a memorandum of agreement was drawn up between the University and the USGS and Ray arrived with his team. Since Ray's recent retirement, the unit is led by Julio Betancourt and Bob Webb with other desert scientists involved including Betsy Pierson who continues the saguaro monitoring. Another inspired member of the group is Jan Bowers, who writes books on her own time. Besides a biography on Forrest Shreve, and the ecological controversies of his day, Jan has introduced her readers to a biogeography of desert mountains (*The*

Mountains Next Door), the stark beauty of desert dunes (*Seasons of the Wind*) and my favorite on her creative approach to gardening in Tucson with its wonderful slant on Jan's own changing views: *A Full Life in a Small Place*. The arrival of the USGS group also on the Hill ensured some new work on fossil middens. In addition, Bob Webb and Alex McCord innovated the dating of floods by tree-ring dating of scars on riparian trees. Bob undertook the repeat photographic analysis of Stanton's 100-year-old views of Inner Gorge vegetation in the Grand Canyon, detecting expansion of frost sensitive plants. Poor Bob! We all pitied his fate, to have to conduct research within the Inner Gorge of the Grand Canyon on innumerable river trips.

Besides mischievous ring tails, the grounds of the Desert Laboratory feature interesting animals such as native mule deer, a herd or two of resident javelina (collared peccary), foxes, cottontails, and an occasional bobcat. In season, the Gila monster and three species of rattlesnakes may be expected on or near walkways, to the delight of ophidiophiles and the horror of ophidiophobes. Occasionally a rattler slips indoors, including the day I reached for my telephone, carelessly left on the floor, and discovered a hatchling rattlesnake coiled in the finger well. A photograph of the rattler in the telephone appeared above the masthead of the August 23, 1977, issue of the *Arizona Daily Star*.

Perhaps the most unusual animal on the grounds is an icon, a life-size model of a Shasta ground sloth made of welded wire, and known to all comers on the Hill as "Weldon." Being only an outline, not a fully embodied replica, Weldon is inconspicuous, as seems appropriate for an extinct beast. Weldon haunts a space between our cardon cactus and the South American creosote bush planted in the patio cactus garden years ago by ecologist Tien Wei Yang.

In the summer of 1977 Weldon was fleshed out with papier-mâché, covered with brown "fur" and given a lead part in a bit of Tumamoc theater. It began when I received a copy of a one-act play, *Range Wars*, recently scripted by the Bay Area poet, Michael McClure. This was soon after Dick Felger brought Michael up to the Hill for a visit, and I gave him some reprints. One hardly hopes for such a dramatic response to reprints. From *Science* Michael had incorporated my account of what happened to ground sloths (and other Late Pleistocene megafauna) into an anachronistic scramble of Lincoln County, New Mexico, characters including Billy the Kid and his girl friend, and three half-naked prehistoric ground sloth hunters. Weldon would be speared and doused with ketchup in our version of the play. The dramatis personae included Marty Eberhardt, Gary Nabhan, Karen Reichardt, Sandra Turner and Sandra's tenant at the time whose name was Billy. Naturally he was type cast as "the Kid."

Lacking a stage, we held forth in the open between buildings, the actors appearing in a spotlight in front of the audience on chairs in the parking lot. Mahina Drees, assisted by Tony Burgess and Steve McGlaughlin, directed the world premier (so Michael McClure assured us) to a packed parking lot, two nights running. All members of the troupe, except Billy, who was a real estate appraiser in real life, went on to distinguish themselves in ethnobiology, ecology, botany, or all of the above. A Thespian talent may not be necessary for professional success but I suspect it helps. Weldon is again disembodied in his own private *Walden*, waiting for another chance to go public.

In the mid 1970s, Al Solomon left for Oak Ridge National Laboratories to be replaced by palynologist Vera Markgraf. Vera and her students, including Ben Brown, Faith Duncan, Pat Fall, Bonnie Fine, Sue Fish, Pacifico Payawal and Bob Thompson developed research projects from the Great Basin, México, South America, and the Philippines and were joined by Hector d'Antoni who had won a Guggenheim Fellowship to work at the Desert Lab. Louis Scott of Bloomfontein in South Africa and Gurdip Singh of the Australian National University arrived on sabbatical leaves. Pat Fall published some research which illuminated a serious limitation of alluvial palynology, the tendency for pine pollen to be overrepresented in fine-grained sediments.

In 1978 Dave Steadman came to the Desert Lab with ambitious plans for investigating the prehistoric fossil record on oceanic islands. He expected to find (and did) many avian extinctions, not only in Pacific Islands but also in the Lesser Antilles, far more than were known or predicted at the time. In September on South Cottonwood Lane, the home of Ike and Jean Russell, Mary Kay O'Rourke and I were married surrounded by a small group of friends. Another sabbatical leave could be requested soon (they are not automatic). The obvious place to go would be Australia and parts of the Pacific. While Australia would be an ideal place to explore by Cessna, Ike's flying days were over and Jean needed to stay home to care for him in the little time he had left.

Sabbatical III. Australia and the South Pacific

Like other good things, sabbaticals do not come free. No matter how good one's teaching in the six years between eligibility, administrators do not want to grant a leave with pay to some deadbeat to waste on the golf course. I had learned that they especially like to be promised a book. Since I already had one in mind, the promise was easy. Now out-of-date and out-of-print *Pleistocene Extinctions* (1967) was due to be revised. More than that, it needed to be redone with better coverage of a crucial part of the globe where the chronology and pattern of prehistoric extinctions invited close comparisons with America and the Old World.

A menagerie of "giant" marsupials, most larger than the largest living one, the red kangaroo, vanished from Australia in the late Pleistocene. Then much later, within the last 1000 years or so, a dozen giant flightless birds, the moas, disappeared from New Zealand. Other extinct birds such as flightless geese and flightless ibis were being described in Hawaii. Virtually everyone agreed that the moa extinction and all other extinctions on oceanic islands were triggered by the side effects of human colonization. The arguments for climatic change focused on Australia itself, with various Australian paleontologists and archaeologists probing all possible causes for the phenomenon. It was especially challenging since nothing like it had been unearthed in older Australian faunas.

I wanted to learn more about the Australian extinction chronology. Although Geof Hope and Jeanette Hope had not encouraged the idea, I wondered if the Australian caves might not harbor dung deposits or at least collagen-rich bones of extinct animals that would be ideal for direct radiocarbon dates. While Jeanette had found a carcass of an extralocal Tasmanian wolf in a cave, in radiocarbon age it was much too young to represent the time of extinction of diprotodon and other Australian megafauna.

When Lucy Cranwell Smith learned that I planned to visit her native land, New Zealand, she helped me get in touch with the experts. These included archaeologists Roger Green, Roger Duff and Atholl Anderson. Through Lucy and Watson Smith, I had already met the Australian palynologists Donald Walker and A.R.H. (Tony) Martin.

Then without warning a renown Australian anthropologist, Richard Wright, and his family showed up at the Desert Lab. He was returning home from a sabbatical in England via the United States. He wanted us to hear the latest results of excavations at a newly discovered incredibly rich bone bed of extinct giant macropod kangaroos at Lancefield in New South Wales. Like the Boney Springs and other mastodon deposits in Missouri, there was archaeological material in uncertain association.

In the fall of 1979, Mary Kay and I flew to Sydney. Richard and his wife Sonia helped us recover from jet lag in their historic home on Sydney Harbor. They sent us on to Canberra where Jeanette Hope was about to start a field excursion to sites I had long heard about and longed to see. These included Lake Mungo with what were then the oldest direct radiocarbon dates on prehistoric Australians, over twice the age of the Clovis culture in the New World, and lunettes of the Willandra Lakes including Jeanette's "Frog Kill Site" where cultural remains over 20,000 years old were associated with bones of mostly small vertebrates.

By the time I was ready to leave Australia and New Zealand four months later I had promises of chapters for a new extinction book from outstanding Australian and New Zealand scientists dealing broadly with the problem of extinctions. And Jeanette Hope was right, there were no cave coprolites or soft tissues to match the fossil organic boluses so common in favorable caves in the Grand Canyon. Ernie Lundelius and others had found *Sarcophilus* coprolite, masses of chewed bone in limestone caves in the Nullarbor Plain. These were not organic. The best Australian dates come from samples of egg shell, on carbon locked into calcite or aragonite in the eggs of extinct birds (bigger than an emu) called *Genyornis*. It was intriguing to note that in paleontological deposits rich in extinct fauna there was no archaeological material in an unquestionable association. Conversely, in well dated stratified archaeological deposits, such as the recently discovered caves of the Franklin River in Tasmania, there were bones of living large mammals. Of course, there was nothing to match a Clovis site. Otherwise the parallels to North America were striking. None of this affected my long-time friend Ernie Lundelius of the University of Texas and his students (now eminent paleontologists themselves), Russ Graham and Holmes Semken, who believed that climate drove the extinctions.

The answer, I felt, would come from oceanic islands. Dave Steadman took Mary Kay O'Rourke and me to see his lava tube excavations in the Galapagos. Unlike Hawaii, New Zealand, the islands of the West Indies and most of those in the Pacific, the Galapagos escaped late prehistoric extinction of native land birds, reptiles and land mammals. Dave's detailed sampling of many rich cave faunas on the Galapagos yielded numbers of bones of virtually all known species of land vertebrates from the islands, including Darwin's finches. With one possible exception, a giant rat not seen historically, all local extinctions known from the Galapagos occurred in the last 200 years. The rate of historic losses is two orders of magnitude greater than anything Dave could find prehistorically. Radiocarbon dates revealed that the giant rat also survived into historic time. No one knows its fate; an indirect effect of human contact is strongly suspected.

Then Dave and I were invited to visit Storrs Olson and Helen James and a cave they were excavating in Maui. Tom Stafford conducted radiocarbon dating on extinct flightless ibis and flightless geese that vanished close to the time of first human colonization over 1000 years ago, significantly older than extinctions in the Galapagos where prehistoric settlement did not occur and prehistoric extinctions are unknown.

Recently Tom and Dave invited me to a cave they were excavating on Eua in the Kingdom of Tonga. Human arrival was marked in their excavation by sudden appearance of charcoal, around 2500 years ago, earlier than prehistoric invasion of New Zealand and Hawaii. Prehistoric voyagers got to Tonga first.

Best of all, the cave on Eua had a fairly rich record of extinct rails, doves, pigeons, parrots and other land birds, back to an Uranium-

Thorium date on limestone of around 65,000 years ago. In all that interval there were no extinctions or obvious faunal turnovers. If climatic change had an affect at all on island bird extinction, it was not to be seen in this record. A similar lack of extinction until humans arrived is now evident in the case of New Zealand's extinct moas. Without human colonization, oceanic islands, large and small, escape intense extinction.

Before leaving Tonga for Tucson, Dave asked if Tom and I might want to see a colony of fruit bats. We did and Dave drove us to a mango orchard next to a few houses. The trees were alive with "flying foxes."

"What's going on, Dave? I thought you said all the islanders love to eat bats. What spares these guys?"

"Tonga has a king," Dave reminded us. "These are the king's bats. They are taboo. Only he or his agents may kill them."

I could only wish the King of Tonga had been around when the island was first colonized. Royal edict might have saved some of the extinct land birds.

Back in Tucson, Tom Van Devender had been teaching my paleoecology course with his unquenchable enthusiasm. By 1984 *Quaternary Extinctions* was released, winning glowing reviews. That year Jim Mead, Larry Agenbroad, Owen Davis and I also published on the remarkable new Utah cave, Bechan, with its amazing deposit of mammoth and other megaherbivore manures. This was an excellent deposit for determining mammoth diet—mostly grass.

In 1985 with the help of the University's Centennial Fund we held a one-day symposium in Tucson on Quaternary extinction and brought all the participants up to the Desert Lab for an open house. It would be the last time I would see Ed Deevey whose interest in biogeography, geochemistry and radiocarbon dating and pollen analysis inspired my career.

I had one more sabbatical ahead, and one more life to live, as it were, this one to be uncomplicated by the search for fossils. I wanted to collect plants from the Río Mayo in northern México, on the trail of one of the Carnegie Desert Laboratory's last remaining investigators, Howard Scott Gentry.

Sabbatical IV. Plants of the Río Mayo

Thirty years had passed since I had come to Tucson with the intention of taking field trips into northern México. To be sure, there was the excursion to Rancho del Cielo in 1962 and another in 1978. With field classes I had driven the Durango highway in Durango and Sinaloa, crossed the Sierra Madre from Yécora, Sonora, to La Junta, Chihuahua, long before construction of Highway 16 began. In his day Ike Russell flew many of us into tiny airstrips in the Sierra Madre. All his passengers tell stories about bush flying with Ike, one story more hair-raising than the next. The flights into the Sierra became less appealing when hostilities between the "mafioso" (drug dealers) and the Mexican army precluded innocent botanical excursions in light aircraft. We had visited Basaseachic Falls and crossed the Gulf into Baja California to see boojum habitat and whales. Ike piloted Geof Spaulding, Fred Wiseman and me to La Junta, Chihuahua, to find the Chihuahuan species of spruce (*Picea*) endemic to the Sierra Madre Occidental. Fred had organized ambitious trips down into west coast México and developed detailed road logs of the dominant desert plants. Jody Lee Duek, Charlie Drew and I had driven to San Luis Potosi to visit Eric Mellink to see his research plots demonstrating an increase in diversity of native rodents around small scale agricultural plots in the Central Plateau.

All of these trips and many others with scheduled classes to the Sonoran coast, the Yaqui River, or the Magdalena palm canyon were educational. Mexican jaunts were wonderful, romantic, and therapeutic. Anyone who appreciates Richard Shelton's "Mexico" in his book of poetry *The Bus to Veracruz* will understand what I mean.

Mexico is remarkably rich in species. It features far more kinds of rattlesnakes, fence lizards, deer mice, pack rats, pines, oaks, leguminous trees, agaves, and cacti, to name a few groups, than the rest of North America. The daunting diversity fascinates biogeographers and demands an explanation. The ice age fossil record explored by the midden hunters from the Desert Laboratory suggests that México no less than the western U.S. experienced climatic changes and an intrusion of northern species during glacial times. But its richness cannot be explained by climatic change alone. Biogeographers may invest many years, even lifetimes, in studying the biota of México with little to show for by way of theory. Nevertheless, I felt the need for some sort of project beyond class trips or vacation jaunts, wonderful as they were.

Plant explorer Howard Gentry spent many years in the field in México. He viewed the country as a cornucopia, spilling species into other parts of the continent. Arizona gets its fair share, especially in our sky island mountains, which may harbor upwards of 1000 species in one mountain island. Most of the dominant species of plants in the Sonoran Desert and desert grassland of southern Arizona can be traced deep into northern México. There they grow in remarkable combinations with a variety of related species unknown north of the border. Palms are found with pines, fir trees with tank bromeliads, and tropical figs with sugar maples, to name a few.

In the Great Depression Howard Gentry somehow supported himself through sale of natural history specimens while undertaking a botanical inventory of one drainage, the Río Mayo of southern Sonora and western Chihuahua located between Ciudad Obregon and Ciudad Chihuahua. The region was virtually unknown botanically when he began his field work. The local names, uses, and legends of plant species native to the region had not been investigated. Tropical dry forest which Howard called "short tree forest" hugs the base of the Sierra Madre and spills into the lower portion of the barrancas. Except for local removal of mine timber, it was all uncut. The pine forest of the Sierra Madre was virgin timber. As Howard found, the native Americans speaking their own languages could reveal much of their unique ecological knowledge to the interested botanist.

Howard Gentry's book, *Río Mayo Plants*, written while he operated out of the Desert Laboratory on Tumamoc Hill, was a powerful inducement. In the early 80s, I found myself drawn increasingly to the Río Mayo country with students or friends collecting plants. The ones we found were not always included in Howard Gentry's monograph of 1200 vascular species (trees, shrubs, forbs, grasses and ferns). New 1:50,000 scale topographic maps helped us tremendously.

By rights I suppose my fourth sabbatical should have been aimed at the extinctions known to occur on the one continent I had not visited, Asia. I could go to Siberia where new Russian radiocarbon dates suggested a gradual decline of woolly mammoths and other megafauna rather than sudden (catastrophic) extinction seen in America. The difference is fascinating. However, I had come to the University of Arizona to be close to México and apart from class trips into Sonora to introduce students to the edge of the tropics I had not attempted any real research in México for many years. Now time was running out. The Río Mayo in southern Sonora was only a day's drive from Tucson. For no more than the promise of adventure itself I could tempt my friends, many of them students or former students, to join a week-long foray into parts of the Río Mayo region which Howard Gentry had not visited. Even if he had, the seasonal changes were so great that we were sure to find some plants in flower that he would have missed. We made eight trips in 1986 alone and over 20 in the next five years.

Members of the field party were invited to collect what attracted their interest, which is what I did. Since uninteresting or drab species like grasses would be neglected, we would spend a few hours collecting nothing but grasses if the season was right. Gentry listed 50 grass species; we now have over 200 from the same region. Some collaborators found special interests. Cathy (Kik) Moore saved big-leaved oaks and worked out their relationships to hydrothermal alteration (acid barrens). Chris Eastoe specialized in a group rich in species in the headwaters of the Río Mayo, the ground orchids. George Ferguson had a sharp eye for pines when he was not discovering new ferns. Jody Lee Duek, Matts Myhrman and Janet Miller filled plastic bags with hard-to-reach specimens at the bottom of Basaseachic Falls. Mike Rourke took coverage data on the woods along Highway 16, the only paved road into the Sierra in our region. Phil Jenkins, Renee Rondeau, Jim Malusa, Guy Clothier and Wayne Van Voorhies returned from an exploration of the Río Mayo west of Moris where I am sure no collectors have been before or since. Dr. Georgie Boyer took on the assignment of collecting every weed species in a mountain bean field. Dick Barber and I joined Prof. Richard Spellenberg from New Mexico State University to help him collect oaks. While driving full tilt along lumber truck roads in the Sierra, Rich can spot rare oak hybrids. With his students, Prof. Andy Sanders of the University of California at Riverside has collected the foothills and coastal plain at the Río Mayo . They keep adding records to the region. Tom Van Devender does the same along the Río Cuchujaqui, and in the Sierra de Alamos. Angelina (Gela) Martinez video taped the uncut forest of the Upper Río Cuchujaqui threatened by buffel grass clearings. Dave Yetman has found that the ethnobotanical lore of the Mayo and other Indian groups is much richer than even Howard Gentry reported. Some Quaternary experts, archaeologists Dennis and Peggy Stanford, paleontologist Elaine Anderson and anthropologist Don Grayson along with Charlie Drew were lured into a week-long collecting foray ending at a lumber mill in the well-named, rarely-visited Sierra Oscura. Charlie's small station wagon was the first vehicle of its class to get over the miserably rough bouldery track into the Sierra. The indefatigable Howard Gentry had been there 55 years earlier. Deep incisions of the Sierra Oscura remain tierra incognita, at least to plant collectors, if not to the Serranos, the mountain ranchers and Indian inhabitants who frequent paradise.

If there were few fossils of extinct animals in the Rio Mayo country, there were surrogates. Livestock allowed to graze in the woods in natural habitat reveal how plants protected themselves from the extinct natives. These would have included native American species of bison, horse, camel (*Camelops*), mammoth and glyptodonts which once served to disperse the fruits of some of the native trees. Ecologist Dan Janzen and I even managed to publish on some of this in *Science*.

Resurrection

The years 1986 and 1987 saw the beginning of a resurrection at the Desert Lab. The buildings were painted inside and out and new flooring installed. Joe McAuliff served as a one-year replacement to teach my courses and to initiate some very important studies on desert perennials and desert soils, an aspect of geobotany that could only be accomplished when a skilled desert ecologist such as Joe is enmeshed in a department strong in geomorphology such as mine.

Even more valuable to us, Joe contacted unit heads in Ecology, Tree-Ring and Arid Lands and initiated a Desert Laboratory scientific advisory committee to begin to find a suitable niche for us within the University. Long-term research about the desert was our claim to fame. The Arizona-Sonora Desert Museum helped by employing Tom Van Devender as a research scientist and Bob Webb and Julio Betancourt attained positions of leadership with the U.S. Geological Survey on Tumamoc Hill.

Finally, in part stimulated by Larry Agenbroad's success at initiating private fund raising for the Mammoth Site, I decided the Desert Laboratory might try the same approach. There were friends of the Carnegie Botanical Laboratory from years ago still living in Tucson who thrilled to the idea of a rejuvenation. There were others who endorsed the geochronology initiative and the concept of an off-campus "beacon-on-the-hill." Friends and supporters may not know or care about all the details of current research, but they can appreciate the desert itself and the mystery of its comings and goings over the eons. The Laboratory could add value, historically and spiritually as well as analytically to the understanding of what deserts are all about. We only had to invite some of our friends to support us in whatever way they chose. We did and they came forward.

In 1987 we formed an Advisory Committee chaired by Dr. Bud Simons, a recently retired surgeon. We issued a case statement and within a year raised our first $100,000. After many years of dedicated part-time service, Betty Fink who had served all sorts of roles on the Hill, from administrative assistant to secretary, had to leave us for a full-time job, a wrenching loss. Joe McAuliff left for a faculty position at the University of Nevada, Las Vegas. The momentum of his efforts kept us on the track. I retired from the University in 1989, kept my wonderful office on the Hill (with windows onto the desert on three sides) and devoted as much energy as I could to the campaign.

A brilliant geochemist with strong biological interests in dry lands and their origin, Jay Quade, was hired to replace me. Jay set up his lab on the Hill, learning the ropes so he could guide the Desert Lab Scientific Advisory Committee for the Faculty of Science. We redesigned our half-time position to "Program Coordinator" and hired Diane Boyer (great granddaughter of Godfrey Sykes) and later Tony Burgess. Through their efforts and especially that of Jean Russell and other members of the Advisory Committee including Bob Ames, Dave Brown, Bill Fink, Julian Hayden, Stan Krzyzanowski, Hal and Cynthia Linquist, Tad Nichols, Bud Simons, and Lucy Cranwell Smith we campaigned for a fellowship in the name of "Ike" Russell.

I view the post Carnegie research effort, the last forty years, with considerable satisfaction. At the Desert Lab deep history is more than either long-term field studies or short-term paleontology. Armed with radiocarbon dating the researchers on the Hill showed that the dry lands of western United States and northern México experienced a major displacement of their "life zones" in glacial times. With further study we noticed that glacial times were the norm and the interglacial era in which we live now is abnormal or atypical. Saguaros, palo verdes, pinyon, limber pine, ponderosa pine and spruce, most if not all of our trees and shrubs, are now displaced northward or to higher elevations than the places they occupied over most of the last 40,000 years, our window of chronological detail. This simple discovery will, I predict, take decades to sink in. The details can be found in a new book on packrat middens edited by Julio Betancourt, Tom Van Devender and me.

Mammoth, ground sloth, and saber-tooth cat and other extinctions stripped the continent of its most magnificent animals. No plants were lost to our knowledge. This all happened around 11,000 radiocarbon years ago (about 13,000 calendar years ago) without benefit of an asteroid hit or other cosmic accident. Prehistoric human colonists called Clovis hunters are uniquely seen in the fossil record around that time. There is no solid evidence for *Homo sapiens* in America before 12,000 radiocarbon years ago and the circumstance may actually be the explanation. Whether or not it is, the extinctions of mammoth and ground sloths were so important in altering the ecology of the region, not to mention the hemisphere, that they deserve attention in teaching our children about the history of our country.

America was discovered long before Columbus by colonists so well adapted to the land on which they lived that they could survive in late glacial climates of the subarctic tundras of northern Siberia and Alaska. The mammoth and other megafauna they found when they spread south in America resembled in diversity the game plains of Africa. Vegetation and climate were changing when this all happened. Prehistoric people reached southern Arizona before saguaros and palo verde came to occupy our Sonoran Desert. Important as they are (and the investigators at the Desert Lab are just a tiny fraction of the small army of paleoecologists studying ice age paleoclimate), there is no strong evidence that the climatic changes 11,000 years ago were either unique or even less favorable for mammoths and ground sloths than those they and other large beasts soon to go extinct had endured earlier in the Quaternary. Whether or not the extinctions were a cosmic or a cultural accident, or both, we may sell America short if we fail to recognize the potential of the land to support much greater diversity of large carnivores and herbivores than have won usufruct to date.

Other Desert Lab devotees would find other themes than these to illustrate the accomplishments of the program. Some are aware of the vast potential for studies devoted to explaining differences in the structure and pattern of desert vegetation formed by substrate (soil and rock type), a pattern best seen from low flying aircraft. With carefully crafted experiments some show how the exclusion of certain small mammals powerfully affects the life of certain plants within experimental plots. Control plots themselves can change far more than most field naturalists would guess. The research of University of Arizona ecologist Larry Venable and his students show this in their studies of the seed bank of desert annuals on grounds of the Desert Laboratory. The very latest paper in *Science* on middens is by Pete Van de Water, Steve Leavitt and Julio Betancourt. It shows, through changes both in stomatal density and in stable isotope composition of glacial age limber pine, that the glacial atmosphere was depleted in carbon dioxide. Just possibly glacial age plant communities were inhibited by carbon dioxide depletion with trees less dense and woodlands more open than they are now.

So it seems clear to me that there is much more to look forward to. Years ago Robert MacArthur called field stations the "ecologist's telescope." He was on sabbatical on Tumamoc Hill at the time. The University of Arizona once claimed fame as "The Lamp of the Desert," before it was suspected that some might think of "desert" as pejorative. That is certainly not true of those who study deserts. We think they harbor phenomena more mysterious than the stars. And if graduate education in ecology faces a dilemma, as Robert MacArthur (1969) suggested, I think a place like the Desert Laboratory provides a solution, for reasons of scale as well as site and tradition. The importance of the Desert Laboratory is being rediscovered.

ACKNOWLEDGMENTS

Julio Betancourt, Jim I. Mead, Mary Ellen Morbeck, Mary Kay O'Rourke and Betsy Pierson helped editorially; Jo Ann Overs prepared the manuscript. I am indeed grateful for the September 1993 reunion at the Hot Springs Mammoth Site, especially to Larry Agenbroad, Eddy Clay, Jim Mead and Joe Muller.

LITERATURE CITED

Bowers, J.E. (1988). "A Sense of Place: The Life and Work of Forrest Shreve." University of Arizona Press, Tucson.

Burgess, T.L., Bowers, J.E., and Turner, R.M. (1991). Exotic plants at the Desert Laboratory, Tucson, Arizona. *Madroño* **38**, 96-114.

Goldberg, D.E. and Turner, R.M. (1986). Vegetation changes and plant demography in permanent plots in the Sonoran Desert. *Ecology* **67**, 695-712.

Hastings, J.R. and Turner, R.M. (1965). "The Changing Mile." University of Arizona Press, Tucson.

MacArthur, R. (1969). Commentary: the ecologist's telescope. *Ecology* **50**, 353.

Martin, P.S. and Wright, H.E., Jr. (1967). "Pleistocene Extinctions: The Search for a Cause." Yale University Press, New Haven.

McGinnies, W.G. (1981). "Discovering the Desert: Legacy of the Carnegie Desert Botanical Laboratory." University of Arizona Press, Tucson.

Chapter 2

Paul Martin, a Man of the Pleistocene

LARRY D. AGENBROAD

Department of Geology and Quaternary Studies Program, Northern Arizona University, Flagstaff, Arizona 86011; and The Mammoth Site, Hot Springs, South Dakota 57747

We are gathered to present a symposium in honor of Paul Martin. Most of the speakers are Paul's former students.

I was never one of Paul's students. I attended the University of Arizona long enough to gain four degrees, but my orientation in academics at that time did not lead me into Paul's classes. I did attend one of his Tumamoc Hill Graduate Seminars, on a Friday afternoon, at the invitation of one of the students. It was my first encounter with a personality who has been on "center stage" in the unfolding drama of Pleistocene extinctions.

The topic of that seminar presentation was "Pleistocene Overkill." It was a new, exciting, promising line of research into the problem of Pleistocene extinctions. Dr. Martin wanted questions and feedback from those present at the seminar, and after a discussion period, he made a statement that really impressed me. He said (I'll paraphrase) "that he didn't know if the Overkill Hypothesis was right, or wrong; that he really didn't care—what he wanted to accomplish was to get people off their posteriors, and into the field to do some solid research."

It has been more than three decades since the idea was proposed. Paul has succeeded, masterfully, at "stirring the pot" and has created or oriented several generations of researchers focused on the late Pleistocene. Not all are in agreement, which provides a glimpse at another facet of Paul's character. He welcomes, even encourages, opposing ideas. Not only that, he encourages researchers who are seemingly opposed to his model to continue their research, and to please keep him posted on their results.

In my opinion, Paul Martin's model of Pleistocene extinction has generated more productive Quaternary research in the past thirty years than any other single concept. It is an **interdisciplinary** model, which fits Paul's research, teaching, and philosophy. It incorporates **archaeology**—the concept that early American hunters wiped out Pleistocene megafauna; **geology**—the medium which contains the evidence, whether it be archaeological, paleontological, or ecological; **paleontology**—the remains of past plant and animal life; **ecology**—plant community reconstructions from pollen, microhistology, macroflora, packrat middens, or dung of a variety of herbivores; and **paleoclimatology**—reconstruction of temperature and moisture conditions of preexisting environments.

Late Quaternary Environments and Deep History: A Tribute to Paul S. Martin, edited by David W. Steadman and Jim I. Mead. The Mammoth Site of Hot Springs, South Dakota, Inc. Scientific Papers, Volume 3. Hot Springs, South Dakota, 1995.

Paul is a stimulating scientist—one who has made a major contribution to Quaternary science by his own efforts, but also by stimulating other researchers. Paul's approach to Quaternary extinctions was done in a style that yielded productive results on the part of all researchers involved (pro and con). He has a synthetic world view. He finds the entire globe fascinating, whether it be island or continent, hot or cold, wet or dry. He has no geographic hangups.

Paul serves as a Quaternary catalyst, forcing people to approach a problem in a multidisciplinary manner. He pursues the cause of Pleistocene extinction from any and all directions—serving as a rallying factor for all disciplines interested in the subject.

Combining some of my feelings with thoughts from colleagues, selected at random, I will attempt a brief summary of Paul's attributes as a scholar, researcher, teacher, colleague, and person. It will be a meld of feelings expressed by those persons mentioned in the acknowledgments. One contribution, from Dr. Guthrie, I have chosen to include in its entirety:

> "The growing edge of knowledge is mainly constructed by the daub of little questions and the cobbles of little truths. the truths are adopted, becoming the givens, as if we always knew them, but the questioners are forgotten. Now and then comes a person staggering under a big question looking for a big truth. They make such an imprint we never forget, can never forget.
>
> These people crystallize issues, cause us to stop, look up, and see differently, phrase our little questions with more focus, and sort our little truths with meaning. Unfortunately, there are few who enter the dither of the science hive who have a big question and reach for big truths. Yet the few courageous ones leave their mark on us all. What sorts them out from the rest? A confident disregard for the protocols of academe or some deep need to find the ultimate keystone? I don't know. I may not be one of those, but I know one when I meet one. And I have."

Brief Biography of Dr. Paul S. Martin

Born: Allentown, Pennsylvania, in 1929.

Education

1951	B.A. Zoology; Cornell Univ., Ithaca, NY
1953	M.S. Zoology; Univ. of Michigan, Ann Arbor, MI
1956	Ph.D. Zoology; Univ. of Michigan, Ann Arbor, MI

Experience

1955-56	Postdoctorate in Biogeography, Yale Univ.
1956-57	Postdoctorate in Biogeography, Univ. de Montreal
1957-68	Geochronology Laboratory, Univ. of Arizona
1968-Present	Professor (now Emeritus) & Acting Director of the Desert Laboratory, Univ. of Arizona

Awards

National Science Foundation Predoctoral Fellowship 1952-53, 19954-55

H.H. Rackham Postdoctoral Fellowship 1955-56

National Research Council of Canada Postdoctoral Fellowship 1956-57

J.S. Guggenheim Fellowship 1965-66

Professional Offices

Review Editor *Ecology* 1965-70

President, Arizona-Nevada Academy of Science 1969-70

Vice President, Society of American Naturalists 1974-75

Publications

Dr. Martin's publications have spanned a large reading community. His publications appear in a variety of journals, symposia volumes and books. Often, he has had repeat contributions in several leading journals. The following list provides the published works of Dr. Martin from 1952 to 1992. In forty years, he has been a very productive scholar, generating four major books and 75 published papers, as well as countless oral presentations (Fig. 1).

Books

1963 "The Last 10,000 Years: A Fossil Pollen Study of the American Southwest." University of Arizona Press, Tucson, 87 pp. 2nd printing 1970.

1967 "Pleistocene Extinctions: The Search for a Cause" (P.S. Martin and H.E. Wright, Jr., eds.). Yale University Press, New Haven, 440 pp.

1984 "Quaternary Extinctions: A Prehistoric Revolution" (P.S. Martin and R.G. Klein, eds.). University of Arizona Press, Tucson, 892 pp. Reprinted in paperback 1989.

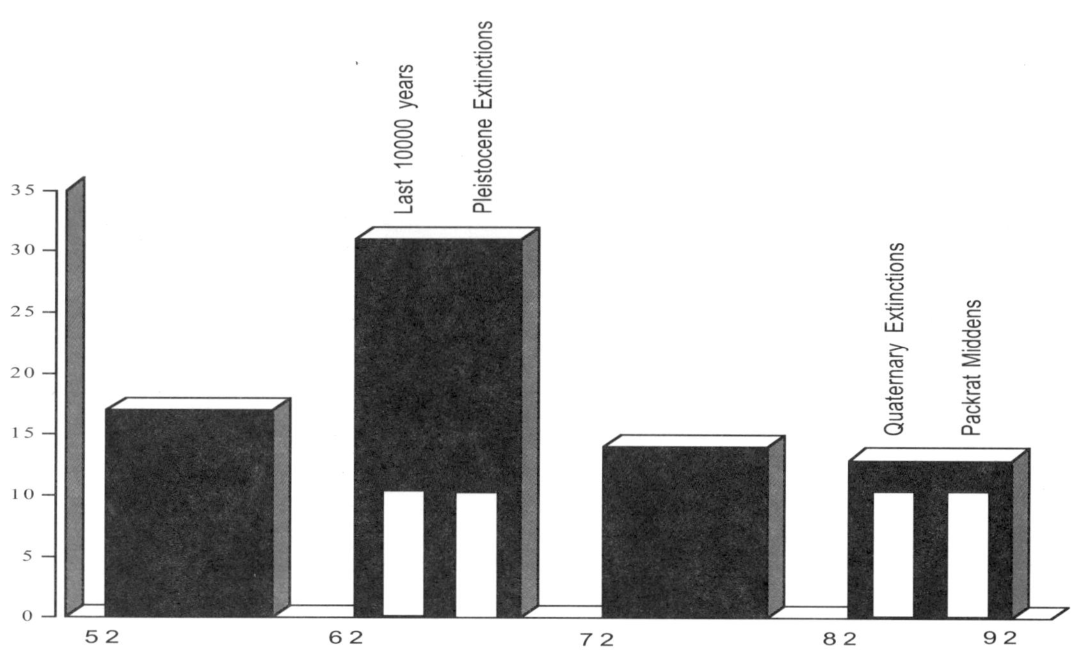

FIG. 1. Numbers of publications by Paul S. Martin grouped by decade.

1990 "Packrat Middens: The Last 40,000 Years of Biotic Change" (J.L. Betancourt, T.R. Van Devender, and P.S. Martin, eds.). University of Arizona Press, Tucson, 467 pp.

Other Publications

1952 A new subspecies of the iguanid lizard *Sceloporus serrifer* from Tamaulipas, Mexico. Occasional Papers of the Museum of Zoology, University of Michigan 543, 1-7.

1954 Birds and biogeography of the Sierra de Tamaulipas (with C.R. Robins and W.B. Heed). *Wilson Bulletin* **66**, 38-57.

1955 Herpetological records from the Gómez Farías region of southwestern Tamaulipas, Mexico. *Copeia* **1955**, 173-180.

Zonal distribution of vertebrates in a Mexican cloud forest. *American Naturalist* **89**, 347-361.

1957 The Pleistocene history of temperate biotas in Mexico and eastern United States (with B.E. Harrell). *Ecology* **38**, 468-480.

1958 A biogeography of reptiles and amphibians in the Gómez Farías region, Tamaulipas, Mexico. Museum of Zoology, Miscellanous Publications 101, 1-102, University of Michigan.

Taiga-tundra and the full-glacial period in Chester County, Pennsylvania. *American Journal of Science* **256**, 470-502.

Pleistocene ecology and biogeography of North America. *In* "Zoogeography," AAAS Symposium Volume, pp. 375-420.

1959 How many logs make a forest. *Ohio Journal of Science* **59**, 221-222.

Mammals from cave deposits in the Gómez Farías region, Tamaulipas, Mexico (with K.P. Koopman). *Journal of Mammalogy* **40**, 1-12.

1960 Arizona's oldest cornfield (with J. Schoenwetter). *Science* **132**, 33-34.

1961 Rampart Cave coprolite and ecology of the Shasta ground sloth (with B.E. Sabels and R. Shutler, Jr.). *American Journal of Science* **259**, 102-127.

Southwestern palynology and prehistory: the last 10,000 years (with J. Schoenwetter and B.C. Arms). Contribution No. 50, Program in Geochronology, University of Arizona, 119 pp.

Geochronology of pluvial Lake Cochise, southern Arizona: I. Pollen analysis of shore deposits (with Richard Hevly). *Journal of the Arizona Academy of Science* **2**, 24-31.

Southwestern animal communities in the Late Pleistocene. *New Mexico Highlands University Bulletin*, pp. 56-66.

Pollen analysis of coprolites. *In* "A survey and Excavation of Caves in Hidalgo County, New Mexico" (M.F. Lambert and J.R. Ambler, Eds.), p. 101-104. The School of American Research Monograph No. 25, Santa Fe.

1962 Pollen analysis and the Cenozoic (with Jane Gray). *Science* **137**, 103-111.

1963 Early Man in Arizona, the pollen evidence. *American Antiquity* **29**, 67-73.

Geochronology of pluvial Lake Cochise. II. Pollen analysis of a 400-meter core. *Ecology* **44**, 436-444.

1964 Detection of fresh and fossil pollen with fluorochromes (with S.J. Shellhorn and H.M. Hull). *Nature* **202**, 315-316.

Paleoclimatology and a tropical pollen profile. *In* "Report of the VIth International Congress on the Quaternary Vol. II: Paleoclimatological Section," pp. 319-323. Lodz, Poland.

Pollen analysis and the full-glacial landscape. *In* "The Reconstruction of Past Environments (J. J. Hester and J. Schoenwetter, assemblers), Fort Burgwin Research Center, pp. 66-73. Taos, New Mexico, Publication No. 3.

Pollen analysis of prehistoric human feces: a new approach to ethnobotany (with F.W. Sharrock). *American Antiquity* **30**, 168-180.

New Paris No. 4: a late-glacial cave deposit in Bedford County, Pennsylvania (with John L. Guilday and Allen D. McCrady). *Bulletin of the National Speleological Society* **26**, 121-194.

1965 Pollen analysis in the Glen Canyon. *In* "1962 Excavation, Glen Canyon Area," University of Utah Anthropological Papers No. 73, 176-195.

Pollen and archaeology at Wetherill Mesa (with William Byers). *American Antiquity* **31**, 122-135.

Geochronology of pluvial Lake Cochise. III. Pollen statistics and Pleistocene metastability (with J.E. Mosimann). *American Journal of Science* **263**, 313-358.

Pleistocene pollen analysis and biogeography of the Southwest. (with Peter J. Mehringer, Jr.). *In* "The Quaternary of the United States" (H.E. Wright and D.G. Frey, Eds.), pp. 433-451. Princeton University Press.

1966 Africa and Pleistocene overkill. *Nature* **212**, 339-342.

1967 Overkill at Olduvai Gorge. *Nature* **215**, 212-213.

Pleistocene overkill. *Natural History*, December, 32-38.

Murray Springs, a mid-postglacial pollen record from southern Arizona (with P.J. Mehringer, Jr., and C.V. Haynes). *American Journal of Science* **265**, 786-797.

Paleolithic players on the American stage. *In* "Arctic and Alpine Regions" (J.D. Ives and R. Barry, Eds.), pp. 669-770. Methuen, London.

1968 Empty nitches by the old corral. University of Arizona Arid Lands Research Newsletter No. 29, 4-5.

1969 Wanted: a suitable herbivore ... *Natural History,* February, 35-39.

Scanning electron photomicrographs of southwestern pollen grains (with C.M. Drew). *Journal of the Arizona Academy of Science* **5**, 147-176.

Pollen analysis and the scanning electron microscope. *In* "2nd Annual SEM

Symposium of the Illinois Institute of Technology Research Institute" (Om Johari, Ed.), pp. 91-103. Chicago.

Plant remains in some coprolites from Utah (with E.O. Callen). *American Antiquity* **34**, 329-331.

The chemical plow in the bush. *Ecology* **50**, 949.

1970 Pleistocene nitches for alien animals. *BioScience* **20**, 218-221.

Ecology in the fourth dimension. *In* "Topics in the Study of Life, the BIO Source Book" (Amy Kramer, Ed.), pp. 372-381. Harper and Row, New York.

On ice age behavior. *Journal of the Arizona Academy of Science* **6**, 91, 179.

Additional scanning electron photomicrographs of southwestern pollen grains (with C.M. Drew). *Journal of the Arizona Academy of Science* **6**, 140-161.

1971 Prehistoric overkill (abridged from Pleistocene Extinctions, the Search for a Cause, P.S. Martin and H.E. Wright, Jr., Eds., Yale Univ. Press, 1967, pp. 612-624). *In* "Man's Impact on Environment" (T.R. Detwyler, Ed.), Chapter 45. McGraw-Hill.

Pleistocene overkill. *In* "Omega, Murder of the Ecosystem and Suicide of Man" (P.K. Anderson, ed.), pp. 42-49. Wm. C. Brown Company. Dubuque, Iowa.

Lehner early man-mammoth site (with Emil Haury, Allen M. Solomon, and James E. King). American Association of Statigraphic Palynologists, Field Trip Guide, October 16, 26 pp. Contribution No. 23, Department of Geosciences, University of Arizona.

1973 The discovery of America. *Science* **179**, 969-974.

Further scanning electron photomicrographs of southwestern pollen grains (with A.M. Solomon, J.E. King, and J. Thomas). *Journal of the Arizona Academy of Science* **8**, 135-157.

Ungulate diets in the lower Grand Canyon (with R.M. Hansen). *Journal of Range Management* **26**, 380-381.

1974 Death of American ground sloths (with Austin Long). *Science* **186**, 638-640.

Extinction of the Shasta ground sloth (with A. Long and R.M. Hansen). *Geological Society of America Bulletin* **85**, 1843-1848.

1975 Vanishings, and future of the prairie. *Geosciences and Man* **10**, 39-49.

Simulating overkill by paleoindians (with J.E. Mosimann). *American Scientist* **63**, 304-313.

Pleistocene riches for alien animals. *BioScience* **20**, 218-221.

Sloth droppings. *Natural History,* August-September, 75-81.

1977 Late Pleistocene biotic communities from the Guadalupe Mountains, Culberson County, Texas (with T.R. Van Devender, A.M. Phillips, and W.G. Spaulding). *In* "The Biological Resources of the Chihuahuan Desert"(R.H. Wauer and D.H. Riskind, Eds.), pp. 104-113. National Park Service, Transactions and Proceedings, Series No. 13.

1979 Natural areas of the Mojave-Sonoran Desert region, Arizona and California: biotic themes. Contract report prepared for the National Park Service, United States Department of the Interior, 352 pp.

Ground sloth dung of the Guadalupe Mountains (with W. G. Spaulding), pp. 259-269 *in* Biological Investigations in the Guadalupe Mountains National Park, Texas (H.H. Genoways and R.J. Baker, eds.), National Park Service, Transactions and Proceedings, Series No. 4.

1980 Shasta ground sloth (*Nothrotheriops shantense* Hoffstetter) at Shelter Cave, New Mexico: environment, diet, and extinction (with R.S. Thompson, T.R. Van Devender, T. Foppe, and A. Long). *Quaternary Research* **14**, 360-376.

1981 Beyond *Bos*: is nothing better? *In* "Southwestern Grasslands, Past, Present and Future" (Carl E. Bock and Jane H. Bock, Eds.), pp. 36-61. Appleton-Whittall Research Ranch, Elgin, Arizona.

1982 Neotropical anachronisms: the fruits the gomphotheres ate (with D. Janzen). *Science* **215**,19-27.

The pattern and meaning of Holarctic mammoth extinctions. *In* "Paleoecology of Beringia" (D.M. Hopkins, J.V. Matthews, Jr., C.E. Schweger, and S.B. Young, Eds.), pp. 399-408. Academic Press.

1984 The Pleistocene dung blanket of Bechan Cave, Utah (with O.K. Davis, L.D. Agenbroad, and J.I. Mead). *In* "Contributions in Quaternary Vertebrate Paleontology" (H.H. Genoways and M.R. Dawson, Eds.), pp. 267-282. Special Publication No. 8, Carnegie Museum of Natural History, Pittsburgh, Pennsylvania.

Catastrophic extinctions and late Pleistocene blitzkrieg: two radiocarbon tests. *In* "Extinctions" (M.H. Nitecki, Ed.), pp. 153-189. University of Chicago Press.

1985 Fossil packrat middens and the tandem accelerator mass spectrometer (with T.R. Van Devender, R.S. Thompson, K.L. Cole, A.J.T. Jull, A. Long, L.J. Toolin, and D.J. Donahue). *Nature* **317**, 610-613.

1986 Refuting late Pleistocene extinction models. *In* "Dynamics of Extinction" (K.D. Elliott, Ed.), pp. 106-130. John Wiley & Sons, Inc., New York.

Extinction of Harrington's mountain goat (with J.I. Mead, R.C. Euler, A. Long, A.J.T. Jull, L.J. Toolin, D.J. Donahue, and T.W. Linick). *Proceedings of the National Academy of Sciences USA* **83**, 836-839.

1987 Radiocarbon dates on bones of extinct birds from Hawaii (with H.F. James, T.W. Stafford, Jr., D.W. Steadman, S.L. Olson, A.J.T. Jull, and P.C. McCoy). *Proceedings of the National Academy of Sciences USA* **84**, 2350-2354.

Clovisia the beautiful! *Natural History*

96, 10-13.

Late Quaternary extinctions: the promise of TAMS ^{14}C dating. *In* "Accelerator Mass Spectrometry" (H.E. Gove, A.E. Litherland, and D. Elmore, Eds.), pp. 179-186. North-Holland, Amsterdam.

1990 Who or what destroyed our mammoths? *In* "Megafauna and Man" (L.D. Agenbroad, J.I. Mead, and L.W. Nelson, Eds.), pp. 109-117. Hot Springs, South Dakota.

40,000 Years of Extinction on the "planet of doom." *Palaeogeography, Palaeoclimatology, Palaeoecology* **82**, 187-201.

1992 Treasures of the Sierra Madre. *Journal of the Southwest* **34**, 391-428.

Parasites of the extinct Shasta ground sloth, *Nothrotheriops shastensis*, in Rampart Cave, Arizona (with Gerald D. Schmidt and Donald W. Duszynski). *Journal of Parasitology* **78**, 811-816.

The last entire earth. *Wild Earth* **2**, 29-32.

ACKNOWLEDGEMENTS

I would like to thank the following individuals, whose written and oral remarks helped generate the symposium presentation, and this manuscript honoring Paul S. Martin: Dr. J. Diamond (Univ. California Medical School, Los Angeles); Dr. R. Graham (Illinois State Museum, Sringfield); Dr. D. Guthrie (Univ. Alaska, Fairbanks); Dr. C.V. Haynes, Jr. (Univ. Arizona, Tucson); Dr. E.L. Lundelius, Jr. (Univ. Texas, Austin); Dr. J.I. Mead (Northern Arizona Univ., Flagstaff); Dr. P. Mehringer (Washington State Univ., Pullman); Dr. D. Steadman (New York State Museum, Albany).

Chapter 3

Extinction of Birds on Tropical Pacific Islands

DAVID W. STEADMAN

Biological Survey, New York State Museum, Albany, New York 12230

INTRODUCTION

For more than 35 years now, Paul Martin has developed and refined the "overkill hypothesis," which argues that human activities, not changing climates and habitats, were the primary cause of late Pleistocene extinctions of large mammals. The overkill hypothesis has been an overwhelming success. Some scientists believe it. Many others do not. Perhaps an even larger number are somewhere in between. The good news is that most researchers in all three groups are a little uncomfortable with their positions and therefore continue to seek answers to questions about late Quaternary extinctions across the entire planet. Regardless of preconceptions or results, the many attempts to test the overkill hypothesis have produced an enormous amount of original and innovative paleoenvironmental research. The participants have included zoologists, botanists, archaeologists, paleontologists, stratigraphers, geochemists, geophysicists, and mathematicians.

While the concept of overkill can be evaluated from a global perspective (Martin, 1984, 1990), it has been examined most rigorously for continental faunas, especially large North American mammals. Various chapters in Martin and Klein (1984) characterize the debate of humans versus climate as the primary or sole cause of the extinctions, with some authors preferring a blend of the two agents (also see Owen-Smith, 1987; Stuart, 1991; Burney, 1993b). Causal studies of extinction depend upon accurate chronologies (Meltzer and Mead, 1983; Mead and Meltzer, 1984). Fortunately, the chronology of North American extinctions continues to improve, with AMS (accelerator-mass spectrometer) ^{14}C dating of bone itself being a major development (Stafford *et al.*, 1990).

In North America, the late Pleistocene loss of so many large mammals (various ground sloths, glyptodonts, horses, llamas, cervids, mammoths, mastodons, bears, cats, etc.) resulted in the extinction of various birds as well. Many of these birds were scavengers (storks, teratorns, vultures, condors, eagles, caracaras) whose abundance and diversity of carrion declined dramatically with the extinction of so many large mammals (Steadman and Martin, 1984; Steadman and Miller, 1987; Emslie, 1987, 1990). Much research on late Quaternary continental faunas focuses on mammals rather than birds or other vertebrates or invertebrates. This taxonomic bias shifts on remote oceanic islands, where birds may be the only

Late Quaternary Environments and Deep History: A Tribute to Paul S. Martin, edited by David W. Steadman and Jim I. Mead. The Mammoth Site of Hot Springs, South Dakota, Inc. Scientific Papers, Volume 3. Hot Springs, South Dakota, 1995.

terrestrial vertebrates present.

On islands near continental source areas, such as the West Indies, late Quaternary faunas often include all classes of vertebrates, with birds usually having the most species (Olson, 1978; Pregill *et al.,* in press). Excavation of archaeological sites on many islands has established the stratigraphic and chronologic association of extinct vertebrates with prehistoric cultural materials. On the West Indian island of Antigua, for example, the bones of 17 extirpated species (5 reptiles, 7 birds, and 5 mammals) have been recovered from the Burma Quarry site, the lower levels of which date to 4300 yr B.P. (Steadman *et al.*, 1984; Pregill *et al.*, 1988). Unlike on continents, it seems to easy for most researchers to accept that prehistoric humans may have been involved in the extinction of island species.

The prehistoric extinction of vertebrates on oceanic islands is calibrated by the Holocene fossil record of the Galápagos Islands, in the far eastern equatorial Pacific. By considering only faunas of Holocene rather than Pleistocene age, we can eliminate the potential effects of major climatic changes associated with the glacial to interglacial transition (documented for the Galápagos by Colinvaux, 1972). Unlike most other Pacific island groups, the Galápagos Islands were uninhabited by humans prior to their discovery by Europeans in AD 1535 (Steadman, 1986). Discounting possible prehistoric visits (see Heyerdahl and Skjolsvold, 1956; Heyerdahl, 1963) as ecologically inconsequential, human impact in the Galápagos was confined to the past 460 years and was relatively minor until about AD 1800.

The Holocene fossil record from five islands in the Galápagos comprises about 500,000 bones of reptiles, birds, and mammals (Steadman *et al.*, 1991). More than 90% of these bones pre-date the arrival of humans. They provide an unusual opportunity to estimate the natural or background rate of extinction. Only 0-3 vertebrate populations vanished in 4000-8000 years preceding human arrival. In the 150-300 years since the arrival of people, 21-24 populations were lost on the same five islands. Thus the background (pre-human) rate of extinction in the Galápagos was about two orders of magnitude (~100 times) less than the rate of extinction after human colonization. When undisturbed by humans, the natural processes of dispersal, establishment, and evolution of vertebrates evidently results in a very low rate of extinction in the Galápagos and perhaps most other tropical oceanic islands.

THE CHRONOLOGY AND EXTENT OF EXTINCTION IN OCEANIA

Virtually all islands in the tropical Pacific (Fig. 1) were inhabited prehistorically and only a few had been abandoned before their subsequent "discovery" by Europeans. Whenever the fossil record has been sampled, prehistoric extinctions of birds have been discovered across Oceania (= Melanesia + Micronesia + Polynesia). These losses are attributed to predation by humans and by associated non-native vertebrates (rats, dogs, pigs), and to landscape changes wrought by prehistoric peoples to accomodate agriculture. The most destructive landscape change has been the loss or alteration of indigenous forests through cutting, burning, and the introduction of non-native plants (Kirch, 1983). Such activities have eliminated natural habitats and have rendered the surviving birds even more vulnerable to the non-native mammalian predators.

The losses of birds in Oceania fall into three categories: 1, extinction (loss of all populations of a species); 2, extirpation (loss of a species on one or more islands although other populations of the species survive elsewhere); and 3, reduced population (loss without replacement of individuals from a surviving population on an island). The last category is the most difficult to demonstrate through the prehistoric record.

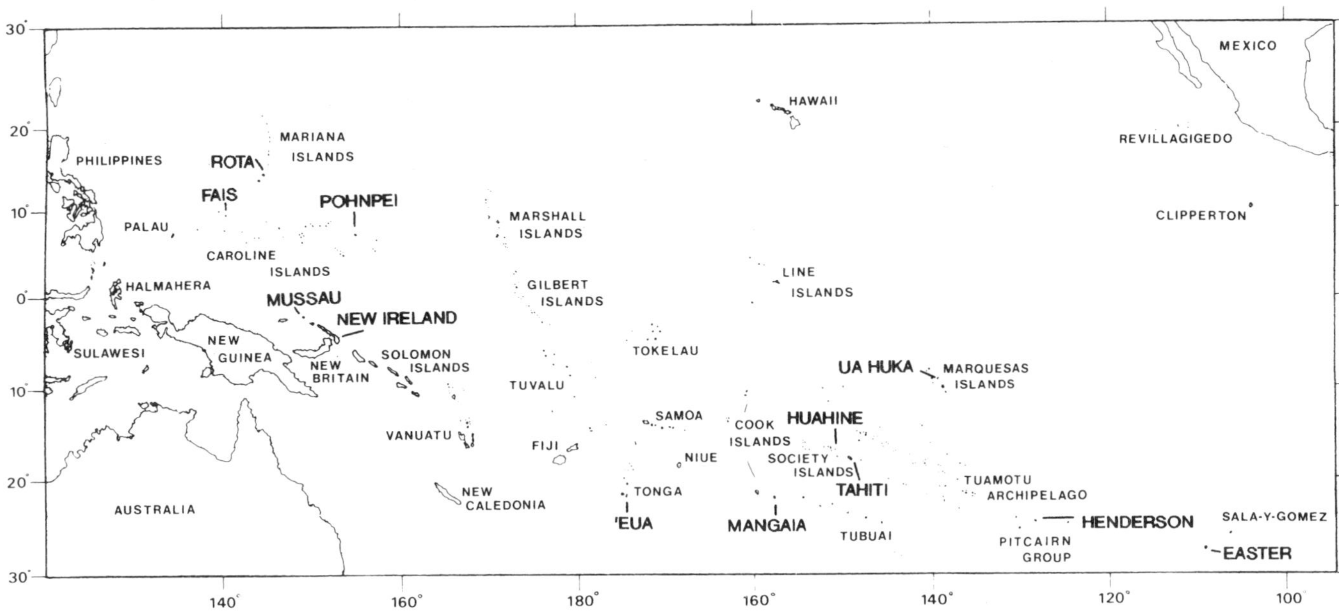

FIG. 1. The Pacific Ocean, showing major island groups. Individual islands mentioned in the text are indicated by arrows.

Micronesia

Bones of native birds have been reported from only three Micronesian islands. On Rota (Mariana Islands), four small cave deposits have yielded bones of 22 species of resident birds, 13 of which are extinct or extirpated (a shearwater, tern, duck, megapode, three rails, two pigeons, a parrot, swift, monarch flycatcher, and parrot-finch; Steadman, 1992a). An archaeological deposit on Fais (Yap) has produced 18 species of resident birds, of which two petrels, two boobies, five terns, a rail, and two pigeons (12 species) are certainly or probably extirpated on Fais today (Steadman and Intoh, 1994). On Pohnpei, very small samples of bird bones from nine archaeological sites have yielded eight resident species, four of which (a shearwater, two petrels, and a megapode) are extirpated (O. Kataoka, W. S. Ayres, and Steadman, unpublished data). I find no cultural, geographical, or biological reason why the record of prehistoric avian extinctions in Micronesia, when more fully studied, will differ from that of Polynesia in anything other than details of taxonomy and chronology.

Melanesia

On average, the islands of Melanesia are larger and sustain richer floras and faunas than those of Polynesia or Micronesia. The prehistoric record of Melanesian birds has been sampled, however, only on Mussau (Steadman, unpublished data), New Ireland (Steadman, J. P. White, and J. Allen, unpublished data), New Caledonia (Balouet and Olson, 1989), and two islands in Fiji (Steadman, 1989b). As in Polynesia and Micronesia, the limited Melanesian record reveals the loss of a great variety of seabirds and landbirds, the former dominated by petrels, the latter dominated by hawks, megapodes, rails, pigeons, parrots, and owls, along with herons, an ibis, button-quail, snipe, owlet-nightjar, and crow. Melanesian populations and species of mammals also were lost, in some cases after prehistoric translocation (Flannery *et al.*, 1988; Flannery and White, 1991).

New Ireland is large (9974 km^2), high (elev. 2399 m), and relatively undisturbed. It currently supports about 108 species of resident landbirds, an avifauna that would seem to be intact. Nevertheless, my preliminary data indicate that even New Ireland has lost about 25% of its species of landbirds since human arrival, including at least two species of flightless rails. New Ireland has been inhabited by people since ca. 33,000 yr B.P. (Allen *et al.*, 1988, 1989; Jones, 1989). In spite of the much longer period of human occupation, a larger percentage of the indigenous avifauna has survived on New Ireland than on Polynesian or Micronesian islands, or on smaller Melanesian islands. This may be due to the buffering effects on human impact of large island size, steep terrain, cold and wet montane climates, and human diseases (such as malaria). Insular species tend to be at greater risk on low, flat, small islands than on high, steep, large islands (MacArthur and Wilson, 1967).

Polynesia: general themes

The prehistoric record of birds is much better known in Polynesia than in Micronesia or Melanesia. In West Polynesia, closer to the more diverse faunas of Melanesia, the prehistoric extinctions have involved a number of reptiles and bats as well as birds (Pregill, 1993; Steadman, 1993a; K. F. Koopman and Steadman, unpublished data). The prehistoric record of Polynesian birds is based upon data collected from archaeological and paleontological sites in the Hawaiian Islands (Olson and James, 1982a, 1982b, 1984, 1991; James and Olson, 1991; James *et al.*, 1987), New Zealand (Anderson, 1984, 1989; Cassells, 1984; Trotter and McCulloch, 1984; Holdaway, 1989; Worthy and Holdaway, 1993), Marquesas Islands (Steadman, 1988, 1989a, 1991a, 1992b; Steadman *et al.*, 1988; Steadman and Zarriello, 1987; Dye and Steadman, 1990), Society Islands (Steadman, 1989a, 1992b; Dye and Steadman, 1990; Steadman and Pahlavan, 1992), Cook Islands (Steadman, 1985, 1987, 1989a, 1991b, 1992b; Allen and Schubel, 1990; Allen and Steadman, 1990; Steadman and Kirch, 1990; Kirch *et al.*, 1991, 1992), Henderson Island (Steadman and Olson, 1985; Schubel and Steadman, 1989), Easter Island (Steadman *et al.*, 1994), Samoa (Steadman, 1993b), Tonga (Steadman, 1989a, 1989b, 1993a; Dye and Steadman, 1990), and Polynesian outliers in Melanesia (Balouet and Olson, 1987; Steadman *et al.*, 1988; Steadman, Pahlavan, and Kirch, 1990). I will review this record briefly, using examples from several island groups.

Most of the deposits are archaeological, representing the last 3500 years or less. The only deposits that are late Pleistocene rather than Holocene in age are from Hawaii, New Zealand, and Tonga. Unlike on continents (Martin, 1990; Stuart, 1991), the Pacific island fossil record shows little if any evidence of vertebrate extinctions at the time of glacial-interglacial transition. Rather, the extinctions occur during the late Holocene, when changes in climate were minor compared to those of the glacial-interglacial transition.

Seabird populations have been depleted across Polynesia since the arrival of people. The loss of seabirds has been particularly severe for petrels and shearwaters, although the distribution and population size of other kinds of tropical seabirds (albatrosses, storm-petrels, tropicbirds, frigatebirds, boobies, terns, gulls) have been reduced as well. The erosion of topsoil associated with deforestation has removed large areas of nesting habitat for burrowing seabirds such as shearwaters and petrels. Species that nest on or within the ground are particularly vulnerable to predation from non-native mammals. Presumably, human navigation across hundreds or thousands of kilometers of ocean was aided inadvertently by various species of seabirds, whose daily and seasonal movements were known to prehistoric voyagers. The loss of most species of seabirds on most islands would have made long-distance voyaging more difficult during late prehistoric times (Steadman, in press).

The losses of landbirds in Polynesia involved all families, although rails, pigeons, doves, and parrots have been the most affected. Unlike East Polynesia, the loss of landbirds in West Polynesia (Tonga, Samoa) includes herons, megapodes, hawks, shrikebills, whistlers, thrushes, white-eyes, and honeyeaters, some of which represent taxa previously known only from Melanesia. By eliminating the "Melanesian" taxa, the anthropogenic extinction of birds on the Tongan island of 'Eua, for example, has artificially magnified the biogeographic distinction between the avifaunas of Polynesia and Melanesia (Steadman, 1993a). Only six of the 27 species of landbirds recorded from pre-human sediments on 'Eua still occur on the island today. Modern distributions alone may not portray the natural biogeographic affinities of island biotas that have been blighted by prehistoric extinctions.

The extinction of so many species of frugivorous, granivorous, and nectarivorous birds, particularly megapodes, pigeons, doves, parrots, and passerines, probably has limited or eliminated the natural means of seed dispersal for certain indigenous Polynesian forest trees. Fruit bats, which also are important in the seed dispersal and pollination of various Polynesian trees (Cox *et al.*, 1991), have suffered prehistoric range contractions as well (Steadman, 1991b). Some of the negative impact people have had on Polynesian birds (and bats) could be reversed with the implementation of rigorous conservation programs (Franklin and Steadman, 1991; Steadman, Greiner, and Wood, 1990). In general, however, it is much too late to restore any true likeness to the natural avifauna of Polynesia.

No family of Pacific birds has lost more species to human impact than the rails, most species of which were flightless. Each island in the tropical Pacific with a suitable fossil record of birds has yielded the bones of one to four endemic species of flightless or nearly flightless rails. At least five genera are involved (*Porzana, Gallirallus, Nesoclopeus, Gallinula [Pareudiastes]*, and *Porphyrio*), with *Gallirallus* providing most of the species discovered thus far. There are ca. 800 islands in Oceania that may have been inhabited by flightless rails at first human contact. Thus the Pacific islands may have harbored upwards of 2000 species of rails that would be feeding, calling, courting, and nesting today in the absence of human colonization. Except for a couple of barely surviving species of *Gallirallus* (Diamond, 1991) and a single species of *Porzana* (Steadman and Olson, 1985; Graves, 1992), we have lost everything but the bones from one of nature's most spectacular examples of adaptive radiation. Imagine what we might be able to study about the adaptive radiation of island species if all those rails still existed.

POLYNESIA: SPECIFIC EXAMPLES

To illustrate the notion of depleted Polynesian avifaunas, I have selected three islands (Ua Huka, Huahine, and Mangaia), in different archipelagos, that have a respectable prehistoric record of birds.

Ua Huka, Marquesas

On Ua Huka (78 km^2, elev. 855 m), Marquesas Islands (Table 1), the Hane site has yielded the largest avian bone assemblage from any tropical Polynesian Island (Steadman, 1989a). Most of these ca. 11,000 bones represent at least seven species of shearwaters and petrels. Eight of the 20 species of seabirds from the Hane site no longer occur on Ua Huka or its offshore islets. Most of the other species of seabirds no longer nest on Ua Huka itself, but only on its tiny offshore islets. Thus the surviving species of seabirds exist in generally reduced numbers and occupy a much smaller breeding range.

The most common landbirds at the Hane site are six species of pigeons and doves, only

TABLE 1. Resident native birds from Ua Huka, Marquesas Islands. Modified from Steadman (1991a). * = extinct species. Modern record = 19th or 20th century specimen or sight record from Ua Huka or its offshore islets.

	Bones from Hane Site	Modern record
SEABIRDS		
Wedge-tailed Shearwater *(Puffinus pacificus)*	x	x
Christmas Shearwater *(Puffinus nativitatis)*	x	-
Audubon's Shearwater *(Puffinus lherminieri)*	x	-
Bulwer's Petrel (*Bulweria* cf. *bulwerii*)	x	x
Tahiti Petrel (*Pterodroma rostrata*)	x	-
Phoenix Petrel (*Pterodroma* cf. *alba*)	x	-
Unknown Petrel (*Pterodroma* small sp.)	x	-
Polynesian Storm-Petrel (*Nesofregetta fuliginosa)*	x	x
White-bellied Storm-Petrel (*Fregetta grallaria)*	x	-
White-tailed Tropicbird (*Phaethon lepturus*)	x	x
Red-footed Booby (*Sula sula)*	x	x
Brown Booby (*Sula leucogaster*)	x	x
Masked Booby (*Sula dactylatra*)	x	-
Abbott's Booby *(Papasula abbottii costelloi)*	x	-
Great Frigatebird (*Fregata minor*)	x	x
Lesser Frigatebird (*Fregata ariel)*	x	x
Gray-backed Tern *(Sterna lunata)*	-	x
Sooty Tern *(Sterna fuscata*)	x	x
Brown Noddy (*Anous stolidus*)	x	x
Black Noddy (*Anous minutus)*	x	x
Blue-gray Noddy (*Procelsterna cerulea*)	-	x
Little Fairy-Tern (*Gygis microrhyncha*)	x	x
LANDBIRDS		
Pacific Reef-Heron (*Egretta sacra*)	x	x
Tuamotu Sandpiper (*Prosobonia* cf. *cancellata*)	x	-
*Undescribed Crake (*Porzana* new sp.)	x	-
*Undescribed Rail (*Gallirallus* new sp.)	x	-
Marquesas Ground-Dove *(Gallicolumba rubescens)*	x	-
*Giant Ground-Dove *(Gallicolumba nui)*	x	-
*Red-moustached Fruit-Dove (*Ptilinopus mercierii)*	x	-
White-capped Fruit-Dove (*Ptilinopus dupetithouarsii)*	x	x
Nuku Hiva Pigeon (*Ducula galeata)*	x	-
*Marquesas Cuckoo-Dove (*Macropygia heana)*	x	-
Marquesas Lorikeet (*Vini ultramarina*)	x	-
*Conquered Lorikeet (*Vini vidivici)*	x	-
*Sinoto's Lorikeet (*Vini sinotoi)*	x	-
Marquesas Swiftlet (*Collocalia ocista)*	-	x
Marquesas Kingfisher *(Halcyon godeffroyi*)	x	-
*Undescribed Monarch (cf. *Myiagra* new sp.)	x	-
Iphis Monarch (*Pomarea iphis*)	-	x
Marquesas Reed-Warbler *(Acrocephalus mendanae*)	-	x
Total species of seabirds	20	14
Combined total species of seabirds	22	
Total species of landbirds	15	5
Combined total species of landbirds	18	

one of which survives on Ua Huka. Three species each of rails and parrots have disappeared as well. While the chronology is in dispute (Kirch, 1986; Spriggs and Anderson, 1993), the oldest portions of the Hane site probably date to ca. 2000 yr B.P. Stratigraphic analysis of avian exploitation at Hane suggests that native birds contributed more than half of all animal protein to the early Marquesan diet, a contribution that was more than halved within about 500 years and declined to insignificance about 500 years later (Dye and Steadman, 1990).

Huahine, Society Islands

The next example is from the Fa'ahia site, Huahine (77 km^2, elev. 669 m), Society Islands (Table 2). Radiocarbon dates from Fa'ahia range from 1140 ± 90 to 770 ± 90 yr B.P. (Steadman and Pahlavan, 1992). All but three of the 15 species of seabirds from Fa'ahia are recorded at the Hane site as well. There is less species-level similarity among landbirds, with only three of Fa'ahia's 15 species shared with Hane. Nevertheless, as at the Hane site, the most common species at Fa'ahia are rails, pigeons, doves, and parrots. A remarkable aspect of the Fa'ahia avifauna is that 30 resident species of birds (only six of which exist on Huahine today) were represented in a sample of only 336 identified bones. A larger bone sample would increase the prehistoric avifauna of the island. Because of the Fa'ahia site, more species of birds are known from Huahine than from the huge (1042 km^2) neighboring island of Tahiti, whose fossil record is unknown.

Mangaia, Cook Islands

The prehistoric exploitation of birds (and other vertebrates) can be evaluated in detail at the Tangatatau Rockshelter (site MAN-44), Mangaia (52 km^2, elev. 169 m), Cook Islands. MAN-44 was excavated in 1989 and 1991 (Steadman and Kirch, 1990; Kirch *et al.*, 1991, 1992, in press; Kirch and Ellison, 1994). The age of this stratified site, based on more than 50 conventional and AMS ^{14}C dates, ranges from about 1400 (zones 1A-1B) to 200 yr B.P. (the protohistoric zone 17). During this period of ca. 1200 years, the human consumption of vertebrates underwent major changes (Tables 3, 4). Domesticates (pig, dog, and chicken) are relatively rare in lower strata but increase upsection as the native landbirds decrease. Pig, dog, and chicken decline in the uppermost prehistoric stratum (zone 15), perhaps because of dietary stress by a relatively large human population. The stratigraphic distribution of bones of Mangaia's only indigenous mammal (the fruit bat *Pteropus tonganus*) resembles that of most extinct birds (see below).

The number of bones from native birds (i.e., all species except chickens) is high in zones 1-4 of MAN-44, relatively low in the middle zones, and high again in zones 15 and 17 (Table 4). The species composition, however, is very different between the early and late zones. Extinct or extirpated species of landbirds dominate zones 1-4, while seabirds, especially the Black-winged Petrel and Audubon's Shearwater, account for the late increase in exploitation of birds. This differs from sites elsewhere in Polynesia, such as the Marquesas, Society Islands, Tonga, or Tikopia, where bones of seabirds are more abundant than those of landbirds in the oldest cultural levels of early sites (Steadman, 1989a, 1989b; Dye and Steadman, 1990; Steadman, Pahlavan, and Kirch, 1990). A factor in the longer survival of certain seabirds on Mangaia may be its precipitous, creviced limestone cliffs, which provide an extensive although narrow band of relatively rat-free nesting habitat for seabirds. It should be noted, however, that only two (Brown Noddy, Common Fairy-Tern) of Mangaia's eight surviving species of seabirds have modern populations in excess of 100 pairs.

Bones of landbirds at MAN-44 are most numerous in zones 1-4, and decrease dramati-

TABLE 2. Resident native birds from Huahine, Society Islands. Modified from Steadman and Pahlavan (1992). * = extinct species. e = recorded in 19th century but now extirpated.

	Bones from Fa'ahia Site	Exists today on Huahine
SEABIRDS		
Wedge-tailed Shearwater (*Puffinus pacificus*)	x	-
Christmas Shearwater (*Puffinus nativitatis*)	x	-
Audubon's Shearwater (*Puffinus lherminieri*)	x	-
Tahiti Petrel (*Pterodroma rostrata*)	x	-
Phoenix Petrel (*Pterodroma alba*)	x	-
Herald Petrel (*Pterodroma arminjoniana)*	x	-
White-tailed Tropicbird (*Phaethon lepturus*)	x	x
Red-footed Booby (*Sula sula*)	x	-
Brown Booby (*Sula leucogaster)*	x	-
Great Frigatebird (*Fregata minor*)	x	-
Lesser Frigatebird (*Fregata ariel*)	x	-
*Undescribed Gull (*Larus* new sp.)	x	-
Brown Noddy (*Anous stolidus*)	x	x
Black Noddy (*Anous minutus*)	-	x
Common Fairy-Tern (*Gygis candida*)	x	x
LANDBIRDS		
Pacific Reef-Heron (*Egretta sacra*)	x	x
Striated [Mangrove] Heron (*Ardeola striata*)	x	-
Gray Duck (*Anas superciliosa*)	-	x
Sooty Crake (*Porzana tabuensis*)	x	-
*Undescribed Rail (*Gallirallus* new sp.)	x	-
Society Island Ground-Dove (*Gallicolumba erythroptera*)	x	-
*Giant Ground-Dove (*Gallicolumba nui*)	x	-
Society Island Fruit-Dove (*Ptilinopus purpuratus*)	x	x
Nuku Hiva Pigeon (*Ducula galeata*)	x	-
Society Islands Pigeon (*Ducula aurorae*)	x	-
*Huahine Cuckoo-Dove (*Macropygia arevarevauupa*)	x	-
Society Islands Lorikeet (*Vini peruviana*)	-	e
*Conquered Lorikeet (*Vini vidivici*)	x	-
*Sinoto's Lorikeet (*Vini sinotoi*)	x	-
Tahiti Swiftlet (*Collocalia leucophaea*)	-	e
Chattering Kingfisher (*Halcyon* cf. *tuta*)	x	x
Society Island Reed-Warbler (*Acrocephalus caffer*)	x	e
Huahine Starling (*Aplonis diluvialis*)	x	-
Total species of seabirds	15	4
Combined total of seabirds	16	
Total species of landbirds	15	4x, 3e
Combined total species of landbirds	18	

TABLE 3. Reptiles and non-human mammals from the Main Excavation Block at Tangatatau Rockshelter (MAN-44), Mangaia, Cook Islands, expressed in NISP. Zone designations (see Figure 6, Kirch et al. in press) as follows: 2-3 includes F57; 4 includes 4A, 4B, 18, F21, F22, F37, F38; 5-7 includes F4, F5, F7, F32; 8 includes F15; 9-14 includes 19, F2, F3, F20, F58; 15 includes F1. No bones were found in zone 16.

Zone	1A	1B	2-3	4	5-7	8	9-14	15	17	Total
REPTILES										
Scincidae/Gekkonidae sp. Lizard	-	-	-	3	-	-	-	1	-	4
Cheloniidae sp. Sea turtle	-	-	-	2	-	1	-	-	1	4
MAMMALS										
Pteropus tonganus Fruit Bat	15	15	33	33	2	-	-	-	3	101
Rattus sp. Rat	1	62	949	252	86	337	301	355	83	2426
Canis domesticus Dog	-	-	35	2	-	5	5	-	4	51
Sus scrofa Pig	-	-	89	23	32	62	23	1	16	246
Delphinidae sp. Porpoise	-	-	-	1	1	-	1	-	2	5
Total NISP (mammals only)										
All species	16	77	1106	311	121	404	330	356	108	2829
All native species	15	15	33	34	3	-	1	-	5	106
All non-native species	1	62	1073	277	118	404	329	356	103	2723

cally by zone 5 (Table 4). Of 17 species of indigenous landbirds, 10 are recorded in zones 5-7 or higher, even if in much reduced numbers. Three of these 10 species (a rail and two parrots) are not recorded above zones 5-7. The most common flightless rail, *Porzana rua,* is last recorded in zone 8. The only extinct/extirpated landbirds recorded above zone 8 are *Gallicolumba* new sp. and the Cook Islands Fruit-Dove *Ptilinopus rarotongensis*. The former species is extinct while the latter still survives elsewhere in the Cook Islands (Rarotonga and Atiu) and is known historically from nearby Ma‘uke and Aitutaki.

As is generally the case in East Polynesia, the 13 species of extinct or extirpated landbirds from MAN-44 consist of rails, sandpipers, pigeons, doves, and parrots. Probably each of the four species of extinct rails were flightless. Their obligatory existence on the ground for foraging and nesting would seem to make them especially vulnerable to predation by humans and rats. Pigeons, doves, and parrots are forest species favored as food by Polynesians. Par-

TABLE 4. Birds from the Main Excavation Block (all excavated units except A25, B25, C25, F10, G40) at Tangatatau Rockshelter (MAN-44), Mangaia, Cook Islands, expressed in NISP. Based on 1989 and 1991 excavations. [Data in Steadman and Kirch (1990) include only 1989 excavations.] Zone designations as in Table 3. * = extant species, extirpated on Mangaia; ** = extinct species. Taxa in brackets [] are not different from others listed more specifically. No bones were found in zone 16.

Zone	1A	1B	2-3	4	5-7	8	9-14	15	17	Total
SEABIRDS										
Puffinus lherminieri	-	-	-	-	-	2	2	10	3	17
Audubon's Shearwater										
**Pterodroma nigripennis*	-	-	-	-	-	19	23	47	11	100
Black-winged Petrel										
**Nesofregetta fuliginosa*	-	1	-	2	3	-	1	1	-	8
Polynesian Storm-Petrel										
Phaethon lepturus	4	2	19	12	2	-	1	1	-	41
White-tailed Tropicbird										
Phaethon rubricauda	-	1	-	-	-	-	-	-	-	1
Red-tailed Tropicbird										
Fregata ariel	-	-	-	-	-	5	-	2	-	7
Lesser Frigatebird										
Fregata minor	-	-	1	-	-	4	-	-	-	5
Greater Frigatebird										
**Sula sula*	-	-	-	1	-	-	-	-	-	1
Red-footed Booby										
Anous stolidus	1	1	-	-	1	5	3	3	1	15
Brown Noddy										
Procelsterna cerulea	-	-	-	1	-	-	-	-	-	1
Blue-gray Noddy										
Gygis candida	2	1	6	2	1	-	-	-	-	12
Common Fairy-Tern										
**Gygis microrhyncha*	-	4	7	1	-	-	-	-	-	12
Lesser Fairy-Tern										
MIGRATORY SHOREBIRDS										
Pluvialis dominica	-	-	2	-	-	-	-	-	-	2
Lesser Golden Plover										
Numenius tahitiensis	5	-	2	2	-	-	-	1	-	10
Bristle-thighed Curlew										
INTRODUCED BIRDS										
Gallus gallus	-	-	22	20	27	16	12	1	2	100
Chicken										
NATIVE LANDBIRDS										
Anas superciliosa	-	-	3	1	2	2	8	11	3	30
Gray Duck										
***Gallirallus ripleyi*	10	5	12	16	1	-	-	-	-	44
Ripley's Rail										
Porzana tabuensis	-	-	2	1	-	1	-	1	1	6
Sooty Crake										
***Porzana rua*	11	44	41	21	2	1	-	-	-	120

Mangaia Crake										
***Porzana* new sp.	-	2	1	1	-	-	-	-	-	4
Tiny Mangaia Crake										
***Porphyrio*? new sp.	-	1	-	-	-	-	-	-	-	1
Small Swamp-Hen?										
***Prosobonia* new sp.	1	1	1	1	-	-	-	-	-	4
Mangaia Sandpiper										
**Gallicolumba erythroptera*	3	5	3	5	-	-	-	-	-	16
Society Island Ground-Dove										
***Gallicolumba* new sp.	-	-	-	-	-	-	7	1	3	11
Medium Ground-Dove										
***Gallicolumba nui*	3	3	-	3	-	-	-	-	-	9
Giant Ground-Dove										
**Ptilinopus rarotongensis*	1	4	4	2	-	-	-	1	-	12
Cook Islands Fruit-Dove										
**Ducula aurorae*	1	2	1	1	-	-	-	-	-	5
Society Island Pigeon										
**Ducula galeata*	3	1	5	2	-	1	-	-	-	12
Nuku Hiva Pigeon										
**Vini kuhlii*	14	41	15	23	1	-	-	-	-	94
Rimatara Lorikeet										
***Vini vidivici*	42	14	3	14	2	-	-	-	-	75
Conquered Lorikeet										
[*/** *Vini kuhlii/vidivici*]	2	2	-	5	-	-	-	-	-	9
Rimatara/Conq. Lorikeet										
Halcyon mangaia	1	2	2	2	-	-	-	-	1	8
Mangaia Kingfisher										
Acrocephalus kerearako	6	6	1	2	-	-	-	-	-	15
Mangaia Reed-Warbler										
Total NISP										
All species	110	143	153	141	42	56	57	80	25	807
All native species	110	143	131	121	15	40	45	79	23	707
Seabirds	7	10	33	19	7	35	30	64	15	220
Migratory shorebirds	5	0	4	2	0	0	0	1	0	12
Native landbirds	98	133	94	100	8	5	15	14	8	475
*/** native landbirds	91	125	86	94	6	2	7	2	3	416
% NISP */** native landbirds										
Of all non-fish	72	57	5	21	4	0.4	2	0.5	2	11
Of all birds	83	87	56	67	14	4	12	2	12	52
Of all native landbirds	93	94	92	94	75	40	47	14	38	88
Total species										
All	16	20	21	23	10	10	8	12	8	32
All native	16	20	20	22	9	9	7	11	7	31
Seabirds	3	6	4	6	4	5	5	6	3	12
Migratory shorebirds	1	0	2	1	0	0	0	1	0	2
Native landbirds	12	14	14	15	5	4	2	4	4	17
Extirpated native landbirds	10	12	10	11	4	2	1	2	1	13

rots (and sometimes pigeons and doves) also were used for their brightly colored feathers (Steadman, in press a).

Four species of resident landbirds recorded from MAN-44 still survive on Mangaia. Each of these species can withstand substantial forest clearance. Because they prefer marshes or dense growths of ferns, grasses, and sedges, the Gray Duck *Anas superciliosa* and Sooty Crake *Porzana tabuensis* probably have benefitted from anthropogenic landscape changes that promote irrigated taro cultivation. The Mangaia Kingfisher *Halcyon mangaia* seems to tolerate moderate forest clearance, although it requires cavities in trees for nesting. Its current scarcity on Mangaia is related to habitat loss and perhaps to nest site competition with the Common Myna *Acridotheres tristis,* an abundant and aggressive Asian bird introduced to Mangaia earlier this century. The Mangaia Reed-Warbler *Acrocephalus kerearako* occurs in forests, thickets, and shrublands of varying disturbance and maturity.

If Tangatatau Rockshelter were Mangaia's only source of paleoenvironmental information, I would be tempted to interpret the distribution of landbird bones (abundant and rich in species in zones 1-4, scarce thereafter) as reflecting a rather rapid (ca. 500 years) decimation of most native species following the first arrival of humans. The problem is the relative scarcity of seabird bones in zones 1-4 (Steadman and Kirch, 1990). In the earliest strata of other "early" East Polynesian sites, the bones of seabirds (dominated by five to ten species of shearwaters and petrels) far outnumber those of landbirds. Zones 1-4 of MAN-44 have not yielded even a single bone of a shearwater or petrel. There seem to be no physical or biological reasons why Mangaia, in pre-human times, should have differed from other Polynesian islands in having a diverse assemblage of shearwaters and petrels.

There is independent evidence that the basal zones at MAN-44 do not represent the initial human presence on Mangaia. Sediment cores extracted by John Flenley and associates in 1989 from Lake Tiriara, 0.9 km from MAN-44, show palynological and stratigraphic signals of human presence on Mangaia beginning at ca. 1600 yr B.P. (Steadman and Kirch, 1990; Kirch *et al.*, 1991, 1992). This sedimentary evidence, all of which relates to forest clearance, features increased influx of clay and charcoal, a decrease in pollen concentration of forest trees, and an increase in the pollen concentration of disturbance indicators (sedges, ferns, non-native species). In 1991, 21 cores from all six major sediment basins on Mangaia were extracted, including Lake Tiriara again (Ellison, 1994). The sediment cores reveal sharp and sustained declines in absolute influx of pollen/spores from all species of trees and ferns characteristic of forests, increases in pollen/spores from species characteristic of disturbed habitats, and the first appearance of charcoal. These data are strong evidence for human presence on Mangaia by ca. 2500 yr B.P., followed by more extensive forest clearance beginning at 1600 yr B.P., this corresponding with the evidence originally recorded in Flenley's cores.

Thus the lake and swamp sediments on Mangaia reveal that major, island-wide human disturbance of the native biota began about two centuries before the earliest cultural deposits at MAN-44. Even more surprising is that, at 1600 yr B.P., Mangaia already had sustained some level of human influence for about 900 years. Unknown is a record of the birds that may have been extirpated during those first 1100 years. Also of interest, and to be discussed briefly below, is how it could be that so many species of birds were able to survive on Mangaia long enough to find their way into the middens at MAN-44.

DISCUSSION

After the 1989 field season on Mangaia, I believed that the rugged *makatea* limestone that covers 56% of the island must have provided a forested refuge for birds during the period preceding deposition of MAN-44 (Steadman and Kirch, 1990). I still believe that this is one of the most important factors that allowed so many species to survive on Mangaia for so long. It is probably no coincidence that the Pacific's only surviving flightless species of *Porzana* is the small rail that inhabits the pinnacle karst of Henderson Island. The most pinnacled, creviced, inhospitable areas of Mangaia's *makatea* still are covered with predominately native forest (Merlin, 1991; Franklin and Merlin, 1992). Nearly all of the sediment deposited in the basins cored by Ellison washed in from the eroded volcanic uplands, not from the *makatea*. Thus the forest disturbance documented in the sediment cores reflects mainly what occurred in the inner volcanic hills, not the *makatea*, where exploitation of natural resources (plants, birds, etc.) probably began in earnest only after the volcanic hills had been overexploited (Steadman and Kirch, 1990).

Another factor that may explain how the birds survived on Mangaia is related to the intensity of island occupation at different times in early prehistory. During the initial period, from 2500 to 1600 yr B.P., there may have been very little or no permanent settlement of the island. Mangaia may have been little more than an occasional outpost for fishermen, perhaps from Rarotonga, who ate seabirds and occasionally set fires during the dry season. Presumably no non-native mammals were introduced. During the next period, from 1600 to 1000 yr B.P., Mangaia may have been visited more frequently and there may have been coastal settlements. Fires increased and the practice of agriculture began. The next period, beginning at 1000 yr B.P., marked the expansion of permanent settlement to the interior of the island, resulting in sites such as MAN-44. This situation may be roughly analagous to the European discovery and settlement of America, with the Vikings representing the Mangaian period before 1000 yr B.P., and Christopher Columbus or the *Mayflower* representing the period after 1000 yr B.P.

The ideas just outlined, along with shoreline changes because of island subsidence and the deep burial of sites by anthropogenic erosion/deposition sequences (Kirch, 1986; Kirch and Ellison, 1994), may help to explain why early habitation sites (those older than 1000 yr B.P.) are so difficult to find in East Polynesia. This difficulty has been the topic of intense debate. The "early group" of archaeologists (Kirch, 1986; Hunt and Holsen, 1991; Irwin, 1992:88-89; Kirch and Ellison, in press) believe that oceanic peoples had discovered much of East Polynesia by 2000 yr B.P. or earlier. These people, however, left behind very few sizeable colonies or habitation sites that have been detected by conventional archaeological methods. Instead, their presence is revealed mainly in "non-cultural" sediment sequences, such as those from the lake and swamp cores from Mangaia. The opposing "late group," most recently represented by Spriggs and Anderson (1993), contends that the scarcity or lack of early East Polynesian sites simply means that people were not present until ca. 1500 to 1000 yr B.P. While I certainly agree that radiocarbon dates should be evaluated critically, several aspects of the hypercritical "chronometric hygiene" put forth by Spriggs and Anderson (1993) may be suspect for geochemical or stratigraphic reasons. My own views, based in large part upon the Mangaian evidence, are in accord with the early group.

Mangaia is not at all unusual in having earlier evidence of human presence in sediment cores than in habitation sites. Similar situations exist for Easter Island (Flenley *et al.*, 1991), Mo'orea (Lepofsky *et al.*, 1992), and even non-

Pacific islands such as Madagascar (Burney, 1993a).

By definition, the chronology of extinction is simply when a species dies out. The timing of the last occurrence may be difficult to determine because it is based in part upon negative evidence; it can be invalidated by even a single younger radiometric date that is geochemically, stratigraphically, and taxonomically unambiguous. Thus any program to study the chronology of extinction should include extensive sampling of the youngest available sediments, thereby reducing the liklihood that a late occurrence has been overlooked. Until a good record of Holocene vertebrates is developed, radiometric dates somewhere in the 10's or 100's of thousands of years ago (such as those reported from Jamaica by MacPhee *et al.*, 1989) may tell us more about the chronology of deposition of bones in caves than about when a particular species became extinct.

After people arrive for the first time on an island, the amount of time for a species to become extinct, whether 200 or 1200 or 2200 years, is an important variable from both a cultural and a conservation perspective. From the standpoint of evolution or biogeography, these differences in timing are less important. Each of these values is a mere blink in geological time, and the finality of extinction is the same in each case. If we were dealing with time scales that exceeded the range of radiocarbon, such millennium differences would be lost within the inherent precision of the dating method.

I have tried to explain how species of birds could survive on islands for many centuries after the arrival of people. I find that explaining survival is more of a challenge than explaining extinction. Extinction is what we have come to expect on islands; survival is the exception. We owe it to the few Polynesian birds that are still around to redouble our efforts toward learning why some species survive and others do not.

Since the 1950's, Paul Martin has tried to convince us that prehistoric human-caused extinctions might be just as feasible on continents as on islands. An abundance of research during the past two decades has confirmed that the activities of prehistoric humans have had profound environmental impacts across the world, whether island or continent. The distinctions between islands and continents get fuzzy when we consider such areas as Great Britain, Australia, New Guinea, Borneo, Sumatra, Madagascar, or even South America. Perhaps for the reasons just explained, the prehistoric extinction of many island birds was not nearly as rapid as the continental blitzkrieg model of Mosimann and Martin (1975) would predict. How long it takes for a species to be obliterated by people depends on a multitude of local factors, among which land area is only one.

ACKNOWLEDGMENTS

This research was inspired by Paul Martin and was funded by the National Geographic Society (grants 2088, 4001-89), National Science Foundation (grants BSR-8607535, BNS-9020750), and the Smithsonian Institution. For research permits, logistics support, and other cooperation, I thank the governments of Easter Island (Chile), French Polynesia, Cook Islands, and Tonga. P. V. Kirch, P. S. Martin, and J. I. Mead kindly commented on the manuscript, which was prepared with the assistance of E. S. Wootan.

REFERENCES

Allen, J., Gosden, C., Jones, R., and White, J. P. (1988). Pleistocene dates for the human occupation of New Ireland, northern Melanesia. *Nature* **331,** 707-709.

Allen, J., Gosden, C., and White, J. P. (1989). Human Pleistocene adaptations in the tropical island Pacific: recent evidence from New Ireland, a greater Australian outlier. *Antiquity* **63**, 548-561.

Allen, M. S., and Schubel, S. E. (1990). Recent archaeological research on Aitutaki, Southern Cooks: the Moturakau Rockshelter. *Journal of the Polynesian Society* **99**, 265-295.

Allen, M. S., and Steadman, D. W. (1990). Excavtions at the Ureia site, Aitutaki, Southern Cook Islands:

preliminary results. *Archaeology in Oceania* **25**, 24-37.

Anderson, A. (1984). The extinction of moa in southern New Zealand. *In* Quaternary Extinctions" (P. S. Martin and R. G. Klein, Eds.), pp. 728-740. University of Arizona Press, Tucson.

Anderson, A. (1989). "Prodigious Birds: Moas and Moa-hunting in Prehistoric New Zealand." Cambridge University Press, Cambridge.

Balouet, J. C., and Olson, S. L. (1987). A new extinct species of giant pigeon (Columbidae: *Ducula*) from archaeological deposits on Wallis (Uvea) Island, South Pacific. *Proceedings of the Biological Society of Washington* **100**, 769-775.

Balouet, J. C., and Olson, S. L. (1989). Fossil birds from late Quaternary deposits in New Caledonia. *Smithsonian Contributions in Zoology* **469**, 1-38.

Burney, D. A. (1993a). Late Holocene environmental changes in arid southwestern Madagascar. *Quaternary Research* **40**, 98-106.

Burney, D. A. (1993b). Recent animal extinctions: recipes for disaster. *American Scientist* **81**, 530-541.

Cassells, R. (1984). The role of prehistoric man in the faunal extinctions of New Zealand and other Pacific islands. *In* "Quaternary Extinctions" (P. S. Martin and R. G. Klein, Eds.), pp. 741-767. University of Arizona Press, Tucson.

Colinvaux, P. A. (1972). Climate and the Galápagos Islands. *Nature (London)* **240**, 17-20.

Cox, P. A., T. Elmqvist, E. D. Pierson, and W. E. Rainey. (1991). Flying foxes as strong interactors in South Pacific island ecosystems: a conservation hypothesis. *Conservation Biology* **5**, 448-454.

Diamond, J. M. (1991). A new species of rail from the Solomon Islands and convergent evolution of insular flightlessness. *Auk* **108**, 461-470.

Dye, T., and Steadman, D. W. (1990). Polynesian ancestors and their animal world. *American Scientist* **78**, 209-217.

Ellison, J. C. (1994). Palaeo-lake and swamp stratigraphic records of Holocene vegetation and sea-level changes, Mangaia, Cook Islands. *Pacific Science 48, 1-15.*

Emslie, S. D. (1987). Age and diet of fossil California Condors in Grand Canyon, Arizona. *Science* **237**, 768-770.

Emslie, S. D. (1990). Additional ^{14}C dates on fossil California Condor. *National Geographic Research* **6**, 134-135.

Flannery, T. F., and White, J. P. (1991). Animal Translocation. *National Geographic Research & Exploration* **7**, 96-113.

Flannery, T. F., Kirch, P. V., Specht, J., and Spriggs, M. (1988). Holocene mammal faunas from archaeological sites in island Melanesia. *Archaeology in Oceania* **23**, 89-94.

Flenley, J. R., King, S. M., Jackson, J., Chew, C. Teller, J. T., and Prentice, M. E. (1991). The Late Quaternary vegetational and climatic history of Easter Island. *Journal of Quaternary Science* **6**, 85-115.

Franklin, J., and Merlin, M. (1992). Species environment patterns of forest vegetation on the uplifted reef limestone of Atiu, Mitiaro, and Ma'uke, Cook Islands. *Journal of Vegetation Science* **3**, 3-14.

Franklin, J., and Steadman, D. W. (1991). The potential for conservation of Polynesian birds through habitat mapping and species translocation. *Conservation Biology* **5**, 506-521.

Graves, G. R. (1992). The endemic land birds of Henderson Island, southeastern Polynesia: notes on nat-natural history and conservation. *Wilson Bulletin* **104**, 32-43.

Heyerdahl, T. (1963). Archaeology in the Galápagos Islands. *Occasional Papers of the California Academy of Sciences* **44**, 45-51.

Heyerdahl, T. and A. Skjolsvold. (1956). Archaeological evidence of pre-Spanish visits to the Galápagos Islands. *Memoirs of the Society for American Archaeology* **12**, 1-71.

Holdaway, R. N. (1989). New Zealand's pre-human avifauna and its vulnerability. *New Zealand Journal of Ecology* **12 (suppl.)**, 11-25.

Hunt, T. L., and Holsen, R. M. (1991). An early radiocarbon chronology for the Hawaiian Islands: a preliminary analysis. *Asian Perspectives* **30**, 147-161.

Irwin, G. (1992). "The Prehistoric Exploration and Colonization of the Pacific." Cambridge University Press, Cambridge.

James, H. F., and Olson, S. L. (1991). Descriptions of thirty-two species of birds from the Hawaiian Islands: part II. Passeriformes. *Ornithological Monographs* **46**, 1-88.

James, H. F., Stafford, T. W., Jr., Steadman, D. W., Olson, S. L., Martin, P. S., Jull, A. J. T., and P. C. McCoy. (1987). Radiocarbon dates on bones of extinct birds from Hawaii. *Proceedings of the National Academy of Sciences USA* **84**, 2350-2354.

Jones, R. (1989). Island occupation. *Nature* **337**, 605-606.

Kirch, P. V. (1983). Man's role in modifying tropical and subtropical Polynesian ecosystems. *Archaeology in Oceania* **18**, 26-31.

Kirch, P. V. (1986). Rethinking East Polynesian prehistory. *Journal of the Polynesian Society* **95**, 9-40.

Kirch, P. V, and Ellison, J. (1994). Paleoenvironmental evidence for human colonization of remote oceanic islands. *Antiquity* **68**, 310-321.

Kirch, P. V., Flenley, J. R., and Steadman, D. W. (1991).

A radiocarbon chronology for human-induced environmental change on Mangaia, Southern Cook Islands, Polynesia. *Radiocarbon* **33**, 317-328.

Kirch, P. V., Flenley, J. R., Steadman, D. W., Lamont, F., and Dawson, S. (1992). Prehistoric human impacts on an island ecosystem: Mangaia, Central Polynesia. *National Geographic Research and Ex ploration* **8**, 166-179.

Kirch, P. V., Steadman, D. W., Butler, V., Hather, J., and Weisler, M. (In press). Prehistory and human ecology in Eastern Polynesia: excavations at Tangatatau Rockshelter, Mangaia, Cook Islands. *Archaeology in Oceania.*

Lepofsky, D., Harries, H. C., and Kellum, M. (1992). Early coconuts on Mo'orea Island, French Polynesia. *Journal of the Polynesian Society* **101**, 299-308.

MacArthur, R. H. and E. O. Wilson. (1967). "The theory of island biogeography." Princeton University Press, Princeton, New Jersey.

MacPhee, R. D. E., Ford, D. C., and McFarlane, D. A. (1989). Pre-Wisconsinan mammals from Jamaica and models of Late Quaternary extinction in the Greater Antilles. *Quaternary Research* **31**, 94-106.

Martin, P. S. (1984). Prehistoric overkill: the global model. *In* "Quaternary Extinctions" (P. S. Martin and R. G. Klein, Eds.), pp. 354-403. University of Arizona Press, Tucson.

Martin, P. S. (1990). 40,000 years of extinctions of the "planet of doom."*Palaeogeography, Palaeoclimatology, Palaeoecology* **82**, 187-201.

Martin, P. S. and Klein, R. G., Eds. (1984). "Quaternary Extinctions." University of Arizona Press, Tucson.

Mead, J. I., and Meltzer, D. J. (1984). North American late Quaternary extinctions and the radiocarbon record. *In* "Quaternary Extinctions" (P. S. Martin and R. G. Klein, Eds.), pp. 440-450. University of Arizona Press, Tucson.

Meltzer, D. J., and Mead, J. I. (1983). The timing of late Pleistocene mammalian extinctions in North America. *Quaternary Research* **19**, 130-135.

Merlin, M. D. (1991). Woody vegetation on the raised coral limestone island of Mangaia, southern Cook Islands. *Pacific Science* **45**, 131-151.

Mosimann, J. E., and Martin, P. S. (1975). Simulating overkill by paleoindians. *American Scientist* **63**, 304-313.

Olson, S. L. (1978). A paleontological perspective of West Indian birds and mammals. *Academy of Natural Sciences of Philadelphia Special Publication* **13**, 99-117.

Olson, S. L., and James, H. F. (1982a). Fossil birds from the Hawaiian Islands: evidence for wholesale extinction by man before Western contact. *Science* **217**, 633-635.

Olson, S. L., and James, H. F. (1982b). Prodromus of the fossil avifauna of the Hawaiian Islands. *Smithsonian Contributions in Zoology* **365**, 1-59.

Olson, S. L., and James, H. F. (1984). The role of Polynesians in the extinction of the avifauna of the Hawaiian Islands. *In* "Quaternary Extintions" (P.S. Martin and R. G. Klein, Eds.), pp. 768-780. University of Arizona Press, Tucson.

Olson, S. L., and James, H. F. (1991). Descriptions of thirty-two species of birds from the Hawaiian Islands: part I. Non-passeriformes. *Ornithological Monographs* **45**, 1-88.

Owen-Smith, N. (1987). Pleistocene extinctions: the pivotal role of megaherbivores. *Paleobiology* **13**, 351-362.

Pregill, G. K. (1993). Fossil lizards from the late Quaternary of 'Eua, Tonga. *Pacific Science* **47**, 101-114.

Pregill, G. K., Steadman, D. W., Olson, S. L., and Grady, F. V. (1988). Late Holocene fossil vertebrates from Burma Quarry, Antigua, Lesser Antilles. *Smithsonian Contributions to Zoology* **463,** 1-27.

Pregill, G. K., Steadman, D. W., and Watters, D. R. (In press). Late Quaternary vertebrate faunas of the Lesser Antilles. *Annals of the Carnegie Museum.*

Schubel, S. E., and Steadman, D. W. (1989). More bird bones from Polynesian archaeological sites on Henderson Island, Pitcairn Group, South Pacific. *Atoll Research Bulletin* **325**, 1-18.

Spriggs, M., and Anderson, A. (1993). Late colonization of East Polynesia. *Antiquity* **67**, 200-217.

Stafford, T. W., Jr., Hare, P. E., Currie, L. A., Jull, A. J. T., and Donahue, D. J. (1990). Accelerator radiocarbon dating at the molecular level. *Journal of Archaeological Science* **18,** 35-72.

Steadman, D. W. (1985). Fossil birds from Mangaia, southern Cook Islands. *Bulletin of the British Ornithologists' Club* **105**, 58-66.

Steadman, D. W. (1986). Holocene vertebrate fossils from Isla Floreana, Galápagos. *Smithsonian Contributions in Zoology* **413**, 1-103.

Steadman, D. W. (1987). Two new species of rails (Aves: Rallidae) from Mangaia, Southern Cook Islands. *Pacific Science* **40**, 27-43.

Steadman, D. W. (1988). A new species of *Porphyrio* (Aves: Rallidae) from archeological sites in the Marquesas Islands. *Proceedings of the Biological Society of Washington* **101**, 162-170.

Steadman, D. W. (1989a). Extinction of birds in East ern Polynesia: a review of the record, and comparisons with other Pacific island groups. *Journal of Archaeological Science* **16**, 177-205.

Steadman, D. W. (1989b). New species and records of birds (Aves: Megapodiidae, Columbidae) from an archaeological site on Lifuka, Tonga. *Proceedings of the Biological Society of Washington* **102**, 537-552.

Steadman, D. W. (1991a). Extinction of species: past, present, and future. *In* "Global Climate Change and Life on Earth" (R.L. Wyman, Ed.), pp. 156-169. Routledge, Chapman and Hall, New York.

Steadman, D. W. (1991b). Extinct and extirpated birds from Aitutaki and Atiu, Southern Cook Islands. *Pacific Science* **45**, 325-347.

Steadman, D. W. (1992a). Extinct and extirpated birds from Rota, Mariana Islands. *Micronesica* **25**, 71-84.

Steadman, D. W. (1992b). New species of *Gallicolumba* and *Macropygia* (Aves: Columbidae) from archaeological sites in Polynesia. *Los Angeles County Museum of Natural History, Science Series* **36**, 329-349.

Steadman, D. W. (1993a). Biogeography of Tongan birds before and after human impact. *Proceedings of the National Academy of Sciences USA* **90**, 818- 822.

Steadman, D. W. (1993b). Birds from the To'aga Site, Ofu, American Samoa: prehistoric loss of seabirds and megapodes. *University of California Archaeological Research Facility Contributions* **51**, 217-228.

Steadman, D. W. (In press). Extinctions of Polynesian birds: reciprocal impacts of birds and people. *In* " Historical Ecology in the Pacific Islands: Prehistoric Environmental and Landscape Change" (T.L. Hunt and P.V. Kirch, Eds). Yale University Press, New Haven.

Steadman, D. W., Greiner, E. C., and Wood, C. S. (1990). Absence of blood parasites in indigenous and introduced birds from the Cook Islands, South Pacific. *Conservation Biology* **4**, 398-404.

Steadman, D. W., and Intoh, M. (1994). Biogeography and prehistoric exploitation of birds from Fais Island, Yap, Federated States of Micronesia. *Pacific Science* **48**, 116-135.

Steadman, D. W., and Kirch, P. V. (1990). Prehistoric extinction of birds on Mangaia, Cook Islands, Polynesia. *Proceedings of the National Academy of Sciences USA* **87**, 9605-9609.

Steadman, D. W., and Martin, P. S. (1984). Extinction of birds in the Late Pleistocene of North America *in* "Quaternary Extinctions" (P. S. Martin, and R. G. Klein, Eds.), pp. 466-477. University of Arizona Press, Tucson.

Steadman, D. W., and Miller, N. G. (1987). California condor associated with spruce-jack pine woodlandin the late Pleistocene of New York. *Quaternary Research* **28**, 415-426.

Steadman, D. W., and Olson, S. L. (1985). Bird remains from an archaeological site on Henderson Island, South Pacific: man-caused extinctions on an uninhabited" island. *Proceedings of the National Academy of Sciences USA* **82**, 6191-6195.

Steadman, D. W., and Pahlavan, D. S. (1992). Prehistoric exploitation and extinction of birds on Huahine, Society Islands, French Polynesia. *Geoarchaeology* **7**, 449-483.

Steadman, D. W., Pahlavan, D. S., and Kirch, P. V. (1990). Extinction, biogeography, and human exploitation of birds on Tikopia and Anuta, Polynesian outliers in the Solomon Islands. *Occasional Papers B. P. Bishop Museum* **30**, 118-153.

Steadman, D. W., Pregill, G. K., and Olson, S. L. (1984). Fossil vertebrates from Antigua, Lesser Antilles: evidence for late Holocene human-caused extinctions in the West Indies. *Proceedings of the National Academy of Sciences USA* **81**, 4448-4451.

Steadman, D. W., Schubel, S. E., and Pahlavan, D. S. (1988). A new subspecies and new records of *Papasula abbotti* (Aves: Sulidae) from archeological sites in the tropical Pacific. *Proceedings of the Biological Society of Washington* **101**, 487-495.

Steadman, D. W., Stafford, T. W., Jr., Donahue, D. J., and Jull, A. J. T. (1991). Chronology of Holocene vertebrate extinction in the Galápagos Islands.*Quaternary Research* **36**, 126-133.

Steadman, D. W., Vargas C., P., and Cristino F., C. (1994). Stratigraphy, chronology, and cultural context of an early faunal assemblage from Easter Island. *Asian Perspectives **16**, 79-96.*

Steadman, D. W., and Zarriello, M. C. (1987). Two new species of parrots (Aves: Psittacidae) from archaeological sites in the Marquesas Islands. *Proceedings of the Biological Society of Washington* **100**, 518-528.

Stuart, A. J. (1991). Mammalian extinctions in the late Pleistocene of northern Eurasia and North America. *Biological Reviews* **66**, 453-562.

Trotter, M. M., and McCulloch, B. (1984). Moas, men, and middens. *In* Quaternary Extinctions" (P. S. Martin and G. G. Klein, Eds.), pp. 708-727. University of Arizona Press, Tucson.

Worthy, T. H. and Holdaway, R. N. (1993). Quaternary fossil faunas from caves in the Punakaiki area, West Coast, South Island, New Zealand. *Journal of the Royal Society of New Zealand* **23**, 147-254.

Chapter 4

Use of Sonoran Rangelands: Lessons from the Pleistocene

ERIC MELLINK

Centro de Investigación Científica y Educación Superior de Ensenada, Apdo. Postal 2732. Ensenada, B.C., México. U.S. mailing: P.O. Box 434844, San Diego, California 92143

INTRODUCTION

A cow raises her head, barely reaching the pods of a mesquite tree in a central Sonora rangeland. She survives at pain in an unsuitable habitat, as she is an animal of open grasslands and savannas, not of thorn-scrub communities that do not offer her an appropiate diet. A nearby group of goats does a little better. Both these animals are part of the mechanism that allows small farmers/ranchers to make a marginal living out of the range. Ironically, the same habitat is profuse in foliage and pods, that are either out of reach or inadequate for the animals on which the local economy is based. Not even the wild creatures consume much of these resources. The largest herbivores, the deer, prefer forbs, and use only a limited amount of the browse on the trees.

Nearby, a wealthy rancher clears 2,000 hectares of "useless" thorn-srub to create adequate habitat for cattle, following current trends in Sonoran ranching. The clearing of shrublands to establish or promote grasslands causes a reduction in the populations of several animals (Mellink and Luévano, unpublished field notes; Germano and Hungerford, 1981) and, perhaps, even the local loss of some. In Sonora this conversion most often includes the establishment of pasturelands of buffel grass (*Cenchrus ciliaris)*, an introduced species. When the buffel grass forms uniform pasturelands it is unsuitable even for grassland species, such as the endangered masked bobwhite (Garza-Salazar *et al*., 1992).

Although extirpation and reduction of populations of native species is currently considered unacceptable, the creation of pasturelands is supported in the name of food production. Indeed, official policies not only permit this transformation, they promote it through the activities of range experimental stations and schools of agriculture.

Despite the rise of cattle to the pedestal of local symbol of progress, the system is deficient. Neither cattle herding in the unmodified rangelands or the production of cattle in converted pasturelands is solving the local problems of welfare and income. Simultaneously, awarness of the effect of these practices on native biota is growing, and will eventually demand changes in the relationship of ranchers with the environment. Moreover, many of the buffel pasturelands of the state currently

Late Quaternary Environments and Deep History: A Tribute to Paul S. Martin, edited by David W. Steadman and Jim I. Mead. The Mammoth Site of Hot Springs, South Dakota, Inc. Scientific Papers, Volume 3. Hot Springs, South Dakota, 1995.

face severe problems with pest insects *Anelonamia* spp. and *Prosapia* spp. (Homoptera: Cercopidae).

Some of us have suggested that including native wildlife into the productive process will result in an increased efficiency in the use of natural resources, without the need of modifying the habitat. However, this idea is not only incompatible with the current Mexican wildlife law, it is also limited because of the few native herbivores available. These few natives contrast with the rich faunas of other semiarid and arid regions of the earth, notably those of eastern Africa.

Many view this poverty as normal, although the fossil record indicates something rather different. Northwest México once supported as rich and varied a fauna as in those other regions, but the sudden demise of most large species at the end of the Pleistocene created the current impoverishment. The cause of this demise has been strongly debated, and no concensus yet exists. However, the evidence available supports, from my viewpoint, Paul Martin's (1973, 1975a, 1984) hypothesis of the incoming humans as the culprits. Evaluation of the pro and against arguments escape the intention of my paper, and are available elsewhere (Martin and Klein, 1984). Rather, I will start from accepting that the lack of complex herbivore comunities in the continent is evolutionarily anomalous.

One by-product of this event is that the plant communities that were encountered by Europeans upon their arrival to the continent, and that are used now as paradigms for conservation and management, were the product of depauperate animal communities (Martin, 1975b; Janzen, 1986). By not having these animal species, a large amount of plant primary production, indeed adapted to be used and disseminated by animals, goes unused.

Moreover, this means that there is a deep, conceptual problem with most of the research devoted to find ways to exploit rangelands more efficiently. The model of a rangeland in good condition that is used is that of an "artificial" system, devoid of large herbivores (Martin, 1975a). So, the strong research efforts devoted to keep grass communities and erradicate certain species whose seeds are favored by their passage through the guts of cattle, seem out of evolutionary place. The dissemination of mesquite and other plants just reestablishes lost ecological processes (Martin, 1975b; Janzen and Martin, 1982).

SEARCHING FOR AN ALTERNATIVE

Paul Martin has indicated in a number of papers (Martin, 1975a, 1975b, 1981) that exotics should be introduced to "...fill the niches" emptied by the dissapearance of megafauna. His arguments often have not been taken seriously or have been dismissed on rather visceral arguments. Because of the effects of many introduced species on the systems in which they have been introduced, biological introductions are considered altogether unacceptable by many people, including many wildlife specialists, environmentalists, and ecologists. Regardless of this opposition, and often prior to it, several game ranchers and farmers have introduced exotic wildlife species with good success. Elsewhere (Mellink, 1989, 1991, 1992), I have reviewed the benefits that can be obtained from such species.

In trying to design introductions of exotics according to ecological rules, I will review some selected Pleistocene faunal assemblages (Lindsay, 1984; Mead *et al.*, 1984; Shaw, 1981; Skinner, 1942; Van Devender *et al.,* 1985), based on current ecological knowledge. Two potential problems might affect the analysis. Our knowledge of any particular fauna might be incomplete. On the other hand, a single deposit might have members of more than one habitat or age. Both cases will affect the interpretation in opposite directions. To reduce the possibilities of this happening, I have selected certain sites, which seem to be relativelly well known, and are restricted to one faunal stage. The sites are in the Sonoran desert. I also include the Golfo de Santa Clara assemblage, although two different terrestrial habitats are represented.

Suitability of exotic species

Arid rangelands are full of plant anachronisms (*sensu* Janzen and Martin, 1982) that coevolved with particular herbivores. The coevolving herbivores are extinct, but other species exist, in other parts of the world, that use similar types of foods, and could occupy their role. To validate this proposal we should test, then, whether certain wild exotic species exploit resources not used by native wild species. Also, to help in the production of goods, the exotic species should use resources not exploited by domestic animals.

Domestication has improved certain animal traits according to the needs of humans. In this process, other characteristics, especially the ability to cope with harsh environments, have been greatly reduced. When used to crop the foliage from shrub communities in arid and semiarid lands, Sonora's most popular domestic animal, the cow, performs poorly. Criollo cattle might do somewhat better, and wild species even better. This might be true even when compared to goats, a much sturdier browser than cattle. That exotic wild species are better suited than domestic species would be a second proposal to test.

A third issue to deal with pertains to the number of species occupying a range. Pleistocene records show a much greater species richness at each individual site, than currently. This modern poverty of species can explain the large amount of unused range resources. As any single animal species is unlikely to use the diverse array of resources available in Sonoran desert scrub communities (known for their great plant species richness), animal communities formed by many species should allow for a greater carrying capacity than communities with few species.

Creation of complex megaherbivore communities

The use of wild species instead of cattle has proven fruitful in Africa (Skovlin, 1971). The economic success of the scheme proposed here in arid America will depend on the confirmation of the three former hypotesis. These are relatively easy to prove or disprove, and indeed are tested, though not formally, by any rancher that introduces exotics. There is an other issue which might be more subtle: How should complex communities be created.

Paul Martin (1981) proposed three levels of faunal introduction, based on taxonomic affinities, to restore lost diversity. This method might be difficult to develop (especially because of changes in local habitat and lack of knowledge of the particular assemblages at each point), and might not be the most desirable for productive purposes, since the extinct animals could have had low reproductive rates (Schüle, 1992). This focus assumes that systems are somehow fixed in terms of the elements that play different roles in the community. In fact, Murdoch (1975) considered that agricultural systems were unstable because they lacked a "...long shared coevolutionary history". However, several studies have shown that species do not behave in fixed assemblages, but that species are rather individualistic (Brown and Kurzius, 1987, 1989). That is, they are present or absent from a particular suitable site regardless of the presence or absence of other non-competing species. This permits the design of communities using animals from different parts of the world, once assemblage rules are known (or believed to be known).

In reviewing the large herbivores in several Pleistocene communities of the Sonoran

Desert (Table 1), I find no evidence of a fixed set of species being present in some sites, and absent in others. They behave as truly individualistic. Therefore, it would seem to be useful to restore animal communities according to general principles of ecological organization, especially since extinction precludes an absolutely accurate reconstruction of *bona fide* Pleistocene animal communities.

The size of animal has a clear bearing on its physiological needs, which, in turn, determine particular sets of demands and feeding capabilities (McNaughton and Georgiadis, 1986; Owen-Smith, 1988). Therefore, animals of different size are likely to have different dietary requirements. Also, some herbivores of similar size have evolved different feeding strategies (Hofmann, 1989; Janis, 1976). This evolutionary complexity might permit the use of various similar-sized species with different diets.

One current opinion on the organization of communities states that their members are spread along a size gradient, allowing them to coexist. This was originally proposed by G. Evelyn Hutchinson (1959) for Coyixidae that exhibited minimum increments of 30% among adjacent species. This value has been used as 50% by others. Although this general rule has been questioned, it has worked adequately for a variety of taxa, from invertebrates to mammals (Brown, 1975; Price and Brown, 1983; Brown and Bowers, 1984; Bowers and Brown, 1982; Wiens, 1984).

TABLE 1. Mammals present in Pleistocene faunas in the Sonoran deserts. *shast.* = *shastensis*, *hest.* = *hesternus*, *hemion.* = *hemionus*, and *Stock.* = *Stockoceros*.

	La Brisca	Dead-man	Papago	Ventana	El Golfo
Megalonyx wheatleyi					X
Nothrotheriops		*shast.*			sp.
Equus conversidens			X		cf.
Equus cf. *complicatus*					X
Equus occidentalis				X	
Equus sp.	large	X			X
Equus tau	cf.		X		
Tapirus sp.				X	X
Dicotyles sp.				X	
Platygonus compressus			X		
Camelops sp.	*hest.* ?		X		X
Hemiauchenia (small)					X
Hemiauchenia cf. *blancoensis*					X
cf. *Palaeolama*					X
Titanotylopus sp.					X
Cervus sp.			X		
Odocoileus sp.	*hemion.*	X		X	X
Capromeryx cf. *minor*	X				
Stockoceros cf. *conklingi*				X	
Stockoceros onusrosagris			X		
Tetrameryx sp.	X				(*Stock.*)
Bison sp.	sp.		*bison*		
Euceratherium sp.		X			
Ovis n. sp.					X
Cuvieronius sp.					X
Mammuthus imperator	sp.				X

African browsing ungulates generally form "...discretely graded series in terms of body size..." (Owen-Smith, 1985). The large mammal community in the Tarangire Game Reserve (Lamprey, 1963) seems to be organized by habitat and size. The topi-zebra-eland-buffalo community of Rukwa Valley (Vesey-Fitzgerald, 1960) is a little more difficult to judge from the information, but could also be interpreted through a size series. In certain

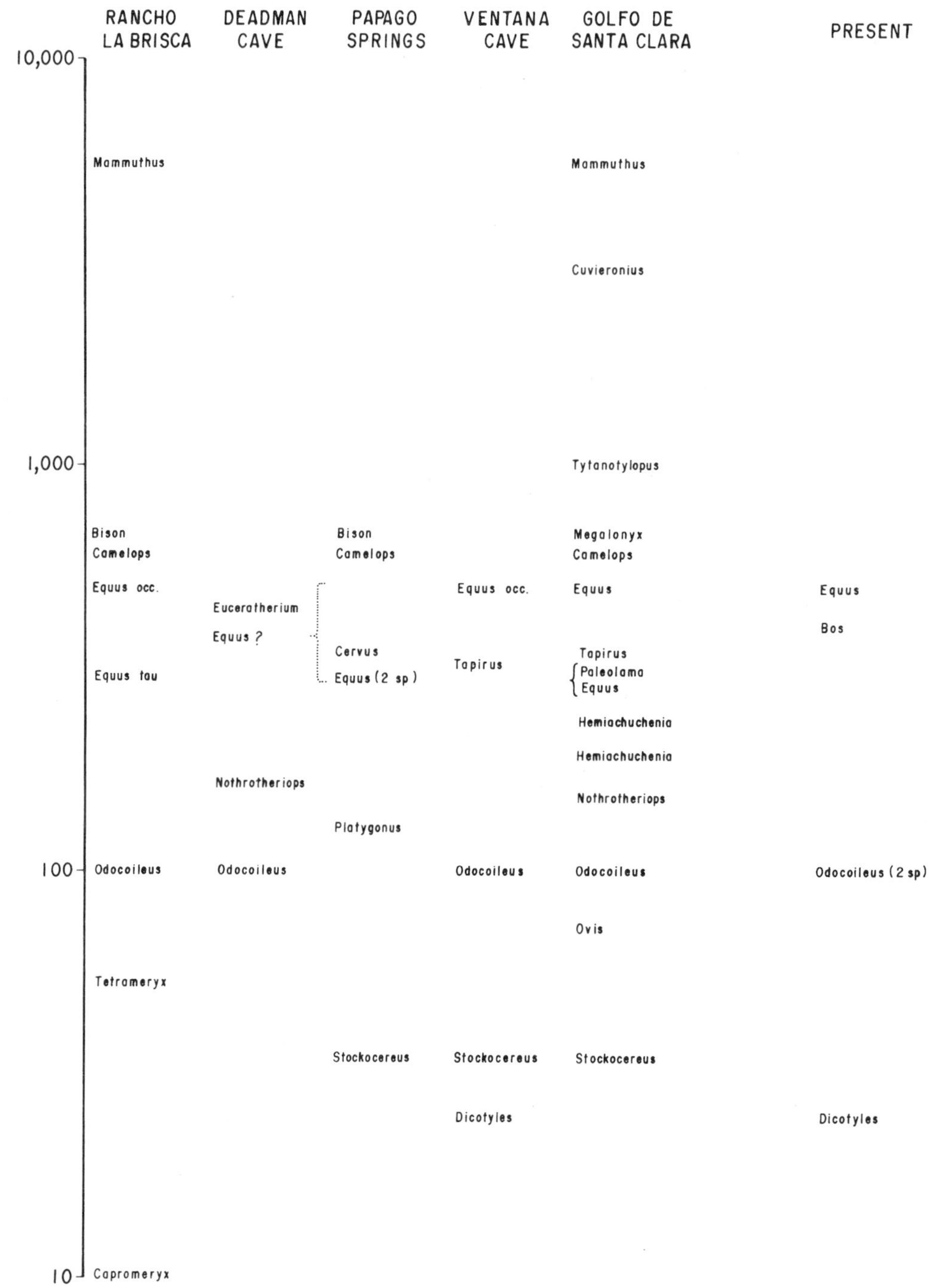

FIG. 1. Size relationships (in kg) among species in Sonoran Desert Pleistocene communities and a current fauna.

African large-herbivore communities the picture is complicated by different species using certain areas at different times of the year (Lamprey, 1963; McNaughton, 1976).

Arrangement of the species in the paleocommunities according to their size shows relatively minor overlap among the species (Fig. 1). Not in all cases is there a minimum 30% increase. The sizes of the species are averaged values from the literature (Anderson, 1984; Kurtén and Anderson, 1980; Mead *et al.*, 1984; Van Devender *et al.*, 1985), that do not include possible geographic variations. The figure does not consider within-site differences in habitat, except that we know that two terrestrial habitats were present in Golfo de Santa Clara. Considering this, the size spread is remarkably in agreement with the theory. The lesson would be: when choosing animals to restock a range, select them along a size gradient, with a minimum of 30% (50% ?) increments in size (providing that they also are from the right habitat). Further, similarly-sized species from different evolutionary lineages might have different diets and be used to enrichen a community.

The system proposed is probably far from perfect. Several traits present in some natural communities are omitted. Some systems exhibit high intensity, low frequency grazing. This pattern is associated with migrations over the range, which would not be possible under the proposed scheme. Although high intensity/low frequency grazing systems can be designed, migratory habits might complicate management.

Several animals migrate at different times (Vesey-Fitzgerald, 1960). Some herbivores make range suitable for other, either by creating clearings or removing lignified forage and stimulating new growth (McNaughton, 1976). This improvement of the habitat for certain species by opening of plant patches or trampling by large animals (Schüle, 1992; Owen-Smith, 1988; Vesey-Fitzgerald, 1960) could have also been a feature of Sonoran Pleistocene ecosystems. Such a patchiness will be difficult to establish, but can be emulated by some form of small, isolated clearings.

Complicating factors for the establishment of complex communities are its ethical considerations, not necesarily concordant with scientific soundness. It is inevitable that changes occur to the range as a result of the introduction of "new" species. However, (1) cattle herding has already caused many changes (see Hastings and Turner, 1965), (2) rangelands necesarily must be kept producing goods for human consumption, and (3) many of the changes proposed herein are designed to restore dormant processes.

Testing the proposal

Exotic megaherbivores have been introduced to arid ranges in North America for different purposes. Game ranches have introduced exotic ungulates to increase the product they offer to the hunters. Other ranches have substituted cattle with one or a few species of large cervids to produce meat. None of these cases is equivalent to the position defended herein. In the first case the production of a huntable surplus requires different management and population sizes than the production of meat. In the second case, the simple substitution of one species (cattle) by another (*e.g.* red deer) does not solve the basic problem of inefficient use of plant growth. It is the creation of multispecific communities that will favor the production of meat, hides and other goods.

Hypotheses on the design of complex systems can be tested in communities of exotic animals introduced in private ranches. These have seldom been studied. Herein I will use the data from a Sonoran ranch where I and my collaborators have studied. In 1985, 242 animals were introduced in a 400 ha plot (subsequently expanded to 800, and later to 1600 ha): 4 giraffe, 2 bison, 6 rhea, 13 pronghorn antelope, 35 blackbuck, 35 barbary sheep, 20 mouflon, 14 eland, 14 oryx, 35 axis deer, 36 Himalayan thar, 16 nilgai, and 12 Pere David's deer. I will try to use these data, supplemented by other sources, to evaluate the different hypothesis on the benefits of introduction of exotic herbivores, and on the structuring of their communities.

1. Do exotic species exploit unused resources ?

On the Dos Arroyos Ranch, southern Arizona, elands strongly preferred browsing over grazing, and consumed very little water (2-3 gal/animal in hot weather, as opposed to 15 gal/animal for cattle) (Biddle, 1976). Camels have been recorded to consume creosotebush (Beal *in* Martin, 1981), a species untouched by cattle and even goats. Goats consume minute quantities of the young growth on limited occasions. Dromedary camels on New Mexico's La Jornada Ranch failed to feed on tarbush, which they were supposed to eradicate. The experimental animals, however, had been accustomed to feeding on alfalfa (Dean Anderson, personal communication).

Analysis of the winter diets of eland and giraffe at the Sonoran ranch show that eland based its diet on legume fruits (17% of identifiable fractions), unidentifiable grasses (17%), ironwood (13%), panicgrass (10%), guayacán (9%), misteltoe and cholla (6% each) and several other plants in lower amounts. It is difficult to evaluate how different this diet is from that of cattle, although cattle prefer grazing. Giraffe consumed mainly ironwood (49%), followed by acacias (14%), misteltoe (11%), legume fruits (9%), and hackberry (5%). Moreover, giraffe feed from the top of the trees, a place unavailable to cattle.

Not all species have different diets than the currently employed domestics, or than the wild natives. Introduced species of deer have caused great concern because they seem to have similar diets to the native deer, and compete with them (Baccus *et al.*, 1985). This is not surprising, since they are closely related species with similar sizes and physiological requirements.

2. Are wild animals better in using Sonoran scrublands than cattle?

Of the species introduced into the Sonoran ranch, Himalayan thar and Pere David's deer (originally from northeastern China) did not do well. Because of their geographic origin, this was not strange. Rheas did not flourish either, the possible reason being consanguineity (Eduardo Gómez, personal communication). The other species have survived without apparent problems and have increased in numbers.

Paul Martin (1981) noted the greater reproductive efficiency of eland and oryx than of cattle. I have no reproductive data on the exotics on the Sonoran ranch. Several problems exist in interpreting the population estimates of 1988 and 1993 (Guillermo Tapia, pers. comm. and Eduardo Gómez, pers. comm., respectively). Removal has taken place; and the carrying capacity for the different species is unknown. Also, reproduction could have been initially low due to newness in the area. Despite shortcomings of the data, growth rate calculated from it might give an idea on how species performed (Table 2). Most species had a larger growth rate during the first period. As these rates derive from sex ratios not too different from 1:1, and include mortality, the figures seem to compare reasonably well with cattle. During the second period the combined

TABLE 2. Annual rates of increase of several species in a ranch in Sonora. See text for further details.

Species	1985-1988	1988-1993
blackbuck	.54	-.15
barbary sheep	.42	-.07
mouflon	.41	?
oryx	.36	.23
eland	.35	.26
axis	.30	-.01
nilgai antelope	.10	.27

effects of a growing population (although the size of the area was also incremented) and mortality, apparently reduced the population growth rates. Nilgai antelopes showed an opposite response, possibly representing an initial need to get used to the area.

3. Is carrying capacity much greater in ranches with multispecific communities?

In the 1600-hectare plot where the introduced species currently live, the range would allow about 67 animal units (a 1000-pound cow with its calf) to be maintained under good condition, and without deteriorating the range (COTECOCA, 1974), in addition to a good number of deer. If criollo cattle were used, or small clearings with grasses were created, as they indeed were, the number would be somewhat larger. Excluding deer, the area supports 356 A.U. in wild animals (based on the average reported size of the different species, transformed to metabolic size, and converted to A.U.). Carrying capacity for the wild species is unknown, and population size for maximum productivity might have been exceeded in some cases. Even so, the standing crop of wild animals greatly exceeds the potential for cattle.

If the area were totally converted to buffel pastureland, it would sustain 400 animal units in cattle (Johnson and Carrillo-Michel, 1976), but at the expense of <u>all</u> native communities. Besides eliminating the vegetation, this would affect severly the native fauna, the hydrographic characteristics of the area, and, perhaps, the potential of maintaining a long-lasting system. Adding the existing deer to the exotic species, 583 A.U. occupy the range. So, under both conditions the metabolic biomass supported by the range is much larger when complex animal communities are used.

4. Is size an adequate criteria to engineer complex communities?

Analysis of the pattern of range use in the Sonoran ranch (Tapia and Mellink, 1989) shows that the species divide in those that use open versus closed habitats, plus some that use the habitats more or less indistinctly. The species in the open habitat system mostly overlap in the areas they use. Contrastingly, the species of the dense habitat (the two deer, thar and barbary sheep) used almost mutually exclusive areas. These differences can be explained on the basis of size and fecal pH (Fig. 2). The open habitat species usually do not overlap in their sizes. Those that did, either occupied different areas (blackbuck and mouflon), had small numbers and very different feeding habits (bison, giraffe, and rhea), or their fecal pH was statistically different (oryx, Pere David's deer, and nilgai antelope). Curiously, the three similar-sized species that coexisted and had different fecal pH come from different parts of the world. Although fine microhabitat differences could be invoked to explain segregation in the species of closed habitats, their catholic habits point to competitive exclusion.

In another study, a community with coexisting horses, goats, and white-tailed deer in the San Luis Potosí Plateau, México showed a clear diet separation (Luévano-Esparza *et al.*, 1993). These two cases seem to prove that arrangement of species along a size gradient, within a certain habitat, can be used for management purposes. They also support that competition between native and exotic deer in Texas should be expected.

CONCLUDING REMARKS

I hope to have shown that establishment of complex communities is a better approach for the use of rangelands, and that exotic species can play a valid and important role. Also, I hope to have provided some guidelines for mangement decisions. I do not claim, however, that exotic species should be introduced widely

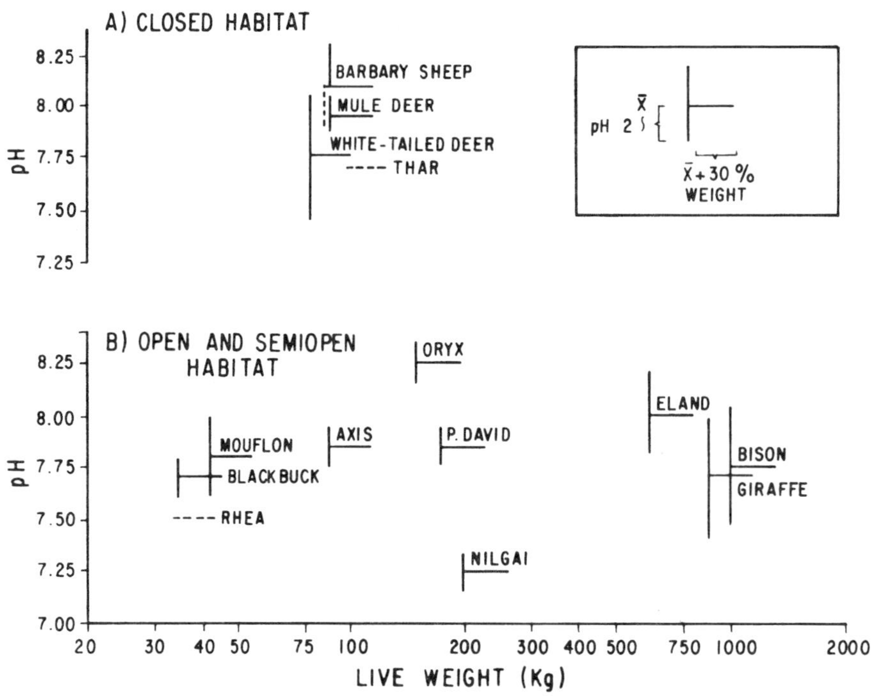

FIG. 2. Weight (kg + 30%) and fecal pH (mean ± 2S.D.) of 14 large herbivores found on a ranch in the Sonoran Desert, México in 1987 (from Mellink, 1991).

and with no control. An important sector of the society has other feelings. Rather, we should intensify studies to support or refute my proposals (and others) and assemblage rules.

ACKNOWLEDGMENTS

This paper could grow out only of Paul Martin's original contributions both on Quaternary extinctions and their effects on plant communities, and on the introduction of exotics. I thank Vicente Ferreira for his review and comments. Guillermo Tapia and Eduardo Gómez kindly supplied data.

REFERENCES

Anderson, E. (1984). Who's who in the Pleistocene: a mammalian bestiary. *In* "Quaternary Extinctions" (P. S. Martin and R. G. Klein, Eds.), pp. 40-89. University of Arizona, Tucson.

Baccus, J. T., Harmel, D. E., and Armstrong, W. E. (1985). Management of exotic deer in conjunction with white-tailed deer. *In* "Game Harvest Management" (S. L. Beasom and S. F. Roberson, Eds.), pp 213-226. Caesar Kleberg Wildlife Research Institute, Kingsville, Texas.

Biddle, J. (1976). Eland studies on Dos Arroyos Ranch. inedit.

Bowers, M. A., and Brown, J. H. (1982). Body size and coexistence in desert rodents: chance or community structure. *Ecology* **63**, 391-400.

Brown, J. H. (1975). Geographical ecology of desert rodents. *In* "Ecology and Evolution of Communities" (M. L. Cody and J. M. Diamond, Eds.), pp. 315-341. Belknap, Cambridge.

Brown, J. H., and Bowers, M. A. (1984). Patterns and processes in three guilds of terrestrial vertebrates. *In* "Ecological Communities: Conceptual Issues and the Evidence" (D. R. Strong, D. Simberloff, L. G. Abele and A. B. Thistle, Eds.), pp. 282-296. Princeton University Press, Princeton.

Brown, J. H., and Kurzius, M. (1987). Composition of desert rodent faunas: combinations of coexisting species. *Annales Zoology Fennici* **24**, 227-237.

Brown, J. H., and Kurzius, M. (1989). Spatial and temporal variation in guilds of North American granivorous desert rodents. *In* "Patterns and the Structure of Mammalian Communities" (D. W. Morris, Z. Abramski, B. J. Fox and M. R. Willing, Eds.), pp. 71-90. Special Publication of the Museum, Texas tech University 28, Lubbock, Texas.

COTECOCA (Comisión Técnico Consultiva para la Determinación Regional de los Coeficientes de Agostadero). (1974). "Coeficientes de agostadero de la República Mexicana; estado de Sonora."

México, D.F.
Garza-Salazar, F. Dobrott, S. J., and Haro-Rodríguez, J. M. E. (1992). Observaciones recientes de las poblaciones de codorniz mascarita *(Colinus virginianus ridgwayi)* en Sonora, México. *Ecológica* **2**, 1-6.
Germano, D. J., and Hungerford, C. R. (1981). Reptile population changes with manipulation of Sonoran desert scrub. *Great Basin Naturalist* **41**, 129-138.
Hastings, J. R., and Turner, R. M. (1965). "The Changing Mile." University of Arizona, Tucson.
Hofmann, R. R. (1989). Evolutionary steps of ecophysiological adaptation and diversification of ruminants: a comparative view of their digestive system. *Oecologia* **78**, 443-457.
Hutchinson, G. E. (1959). Homage to Santa Rosalia, or Why are there so many kinds of animals ? *American Naturalist* **93**, 145-159.
Janis, C. (1976). The evolutionary strategies of the Equidae and the origins of rumen and cecal digestion. *Evolution* **30**, 757-774.
Janzen, D. H. (1986). Chihuahuan Desert nopaleras: defaunated big mammal vegetation. *Annual Review of Ecology and Systematics* **17**, 595-636.
Janzen, D. H., and Martin, P. S. (1982). Neotropical anachronisms: the fruits the gomphotheres ate. *Science* **215**, 19-27.
Johnson, D., and Carrillo-Michel, L. (1976). "Establecimiento de Praderas de Zacate Buffel en el Estado de Sonora." Comité de Fomento Ganadero. Hermosillo, Sonora.
Kurtén, B., and Anderson, E. (1980). "Pleistocene Mammals of North America". Columbia University Press, New York.
Lamprey, H. F. (1963). Ecological separation of the large mammal species in the Tarangire Game Reserve, Tanganyika. East *African Wildlife Journal* **1**, 63-92.
Lindsay, E. H. (1984). Late Cenozoic Mammals from northwestern Mexico. *Journal of Vertebrate Paleontology* **4**, 208-215.
Luévano-Esparza., J., Mellink, E., García-Moya, and Aguirre-Rivera, J. R. (1993). Dieta y traslapo dietario del venado cola blanca, jabalí de collar, cabra y caballo, durante el verano en la Sierra de la Mojonera, Vanegas, S.L.P. *Agrociencia (serie Recursos Naturales Renovables)* **1**(3), 105-122.
Mcnaughton, S. J. (1976). Serengeti migratory wildebeest: facilitation of energy flow by grazing. *Science* **191**, 92-94.
McNaughton, S. J., and Georgiadis, N. J. (1986). Ecology of African grazing and browsing mammals. *Annual Review of Ecology and Systematics* **17**, 39-65.
Martin, P. S. (1973). The discovery of America. *Science* **179**, 969-974.
Martin, P. S. (1975a). Pleistocene niches for alien animals. *BioScience* **20**, 218-221.
Martin, P. S. (1975b). Vanishings, and future, of the prairie. *Geoscience and Man* **10**, 39-49.
Martin, P. S. (1981). Beyond Bos: is nothing better? *In* "Proceedings of Symposium on Southwestern Grasslands: Past, Present and Future" (C. E. Bock and J. H. Bock, Eds.), pp. 33-61. Appleton-Whittell Research Ranch, Elsin, Arizona.
Martin, P. S. (1984). Prehistoric overkill: the global model. *In* "Quaternary Extinctions" (P.S. Martin, and R.G. Klein, Eds.), pp. 354-403. University of Arizona, Tucson.
Martin, P. S., and Klein, R. G., Eds. (1984). "Quaternary Extinctions." University of Arizona, Tucson.
Mead, J. I., Roth., E. L., Van Devender, T. R., and Steadman, D. W. (1984). The Late Wisconsinan vertebrate fauna from Deadman Cave, Southern Arizona. *Transactions of the San Diego Society of Natural History* **20**, 247-276.
Mellink, E. (1989). Megaherbívoros exóticos en la utilización de los agostaderos áridos y semiáridos del norte de México. "Memorias del 7mo Simposio sobre Fauna Silvestre", pp. 110-130. Universidad Nacional Autónoma de México, México, D.F.
Mellink, E. (1991). Exotic herbivores for the utilization of arid and semiarid rangelands of Mexico. *In* "Wildlife Production: Conservation and Sustainable Development" (L. E. Renecker and R. J. Hudson, Eds.), pp. 261-266. Agricultural and Forest Experiment Station. University of Alaska. Fairbanks.
Mellink, E. (1992). Diseño de comunidades complejas de herbívoros. *In* "Primer Simposio Internacional sobre Alternativas y Estrategias en Producción Animal", pp. 318-338. Universidad Autónoma Chapingo, Chapingo, México.
Murdoch, W.W. (1975). Diversity, stability, complexity and pest control. *Journal of Applied Ecology* **12**, 745-807.
Owen-Smith, N. (1985). Niche separation among African ungulates. *In* "Species and speciation" (E. S. Vrba, Ed.), pp. 167-171. Transvaal Museum Monograph 4. Transvaal Museum, Pretoria.
Owen-Smith, N. (1988). "Megaherbivores; the Influence of Very Large Body Size on Ecology." Cambridge University Press, Cambridge.
Price, M. V., and Brown, J. H. (1983). Patterns of morphology andd resource use in North American desert communities. *Great Basin Naturalist Memoirs* **7**, 117-134.
Schüle, W. (1992). Anthropogenic trigger effects on Pleistocene climate ? *Global Ecology and Bioge-*

ography Letters **2**, 33-36.
Shaw, C. A. (1981). "The Middle Pleistocene El Golfo local fauna from northwestern Sonora, Mexico." Unpublished Master's Thesis. California State University, Long Beach.
Skinner, M. F. (1942). The fauna of Papago Springs Cave, Arizona and the study of *Stockocereus*; with three new antilocaprines from Nebraska and Arizona. *Bulletin of the American Museum of Natural History* **80**, 143-220.
Skovlin, J. M. (1971). Ranching in East Africa: a case study. *Journal of Range Management* **24**, 263-270.
Tapia, G., and Mellink, E. (1989). Selección de habitat de trece ungulados y un ave corredora en una reserva en el Desierto Sonorense. *In* "Memorias del 7mo Simposio sobre Fauna Silvestre", pp. 222-241. Universidad Nacional Autónoma de México, México, D.F.
Van Devender, T. R., Rea, A. M., and Smith, M. L. (1985). The Sangamon interglacial vertebrate fauna from Rancho La Brisca, Sonora, Mexico. *Transactions of the San Diego Society of Natural History* **21**, 23-55.
Vesey-Fitzgerald, D. F. (1960). Grazing succession among East African game mammals. *Journal of Mammalogy* **41**, 161-172.
Wiens, J. A. (1984). On understanding a nonequilibrium world: myth and reality in community patterns and processes. *In* "Ecological Communities: Conceptual Issues and the Evidence" (D.R. Strong, D. Simberloff, L.G. Abele and A.B. Thistle, Eds.), pp. 439-457. Princeton University Press, Princeton.

Chapter 5

An Experiment in the Interpretation (Visual Assessment) of Mammalian Bones from Sediments of the Quaternary Period

NIKOLAI K. VERESHCHAGIN

Academy of Sciences, Zoological Institute, St. Petersburg, Russia

While examining faunal remains—bones of fossilized animals—the paleozoologist solves problems of zoology (faunistics, phylogeny, zoogeography, etc.), paleoecology, taphonomy, stratigraphy, geochronology and archaeology. Here everything is closely connected, and the scientist is obliged to be oriented more or less competently (skillfully) in all these subjects.

Having discovered in a geologic stratum a sample—the fragment of some kind of bone—the researcher should quickly learn its skeletal element, species identification, geologic age and the conditions of the animal's death. In short, the researcher should pursue information in a manner similar to Sherlock Holmes.

Having identified during the last half of the century (from 1940) about a million fragments of bones from the Quaternary strata of Northern Eurasia (from river, lake, sea, talus and cave sediments), I want to acquaint the venerable audience with my method of assessing bone material with the help of organoleptics.

Organoleptics is the assessment by our senses (receptors and analyzers) of edible or other products, such as bread, cheese, or meat, industrial material (metals, chemicals, etc.), objects of daily life, and the arts. While organoleptics is carried out by separate specialists, it also is used by all of us widely and often unrealized (instinctively) in our everyday life, while purchasing ordinary products (paintings, clothes, hardware, etc.). Laboratory research of the geologic age of bones, namely potassium-argon, radiocarbon, fluorine, photoluminescence, collagen and others, are usually rather unwieldy; this is already the next stage, the work of the physicist or chemist rather than the paleontologist.

Nevertheless, while working by means of organoleptic methods, the researcher is obliged to have a clear idea about the elementary physico-chemical processes that occur within a bone during its residence in the sediments. The process of fossilization in fresh bone matter is varied. The main constituents of fossil-

Late Quaternary Environments and Deep History: A Tribute to Paul S. Martin, edited by David W. Steadman and Jim I. Mead. The Mammoth Site of Hot Springs, South Dakota, Inc. Scientific Papers, Volume 3. Hot Springs, South Dakota, 1995.

ization are a loss of organics, partial physical destruction (breakage), the solution/dissolution of the calcium base (the skeleton) of bone or its substitution by another mineral (for example by silicon minerals), and changes in color, odor, strength/solidity, specific gravity or other physical qualities. An important aspect is the state of preservation of the bone protein—collagen. Organoleptics is good by the fact that it enables one to assess immediately the whole complex of features, to give information on the animal's death and on the fates of its skeletal remains in the sediment.

It is possible that my American colleagues won't find anything new for their research in what has been said. This is because physical processes and science are extremely similar on our continents.

After this introduction, we will shift to practical examination of osteological specimens from various types of Quaternary deposits.

Here is an outline of the stations/points and areas of our excavations and observations (usually jointly with archaeologists) during the past 50 years.

1. Fluvio-lacustrine (alluvial) burials—hydrogenic.

2. Peat bogs—deposits of overgrown lakes—hydrogenic.

3. Talus slopes of river valleys and hills, and aeolian loess of watersheds.

4. Asphaltic (bitumen), ozokerite deposits and animal traps.

5. Cave deposits of varied types.

6. Bone middens of human settlements in the historical period.

We will examine these different types of deposits according to the numerical sequence.

1. Animals died in valleys of flat rivers from annual floods, and disappeared under the ice of lakes and channels. Their bones of Pleistocene age from banks (beaches) date from the lower to the upper Pleistocene. They are easily distinguished by specific gravity. The lower Pleistocene ones are heavy and saturated with siliceous and ferric salts. The upper ones are considerably lighter and less mineralized secondarily. The bones from beaches are of two taphonomic types.

Type A—from ancient lakes and swamps. Dark, sometimes of coal coloring, with violet tints on the surface—the consequence of coloring by vivianite and manganese.

Type B—from sandy fluvial and deltaic facies. On the surface and in fracture they are the color of a ripe pea. When cut by a scalpel the distinct odor of collagen is sensed. The bones feel sticky to your wet fingers. Their surface is always smooth. Dendritic deposits of manganese and ferric salts are characteristic.

2. Animals, especially forest ungulates, often were killed in or near lakes, swamps, or bogs with floating mats. In Ireland there were many such examples of the death of big-antlered deer (*Megaloceros*). Bones of animals killed in ancient swamps are brown, colored through and through by humic acids, slightly mineralized and relatively light in weight. Those that occur from an ancient sapropel are heavier, as for example the bones from layers of the Forest Bed at Kromer and Norfolk, Northern England. Bones from layers of peat and sapropel of the Neolithic epoch (3-4 thousand years old) in the region near the Baltic Sea are light brown. These layers represent the settlements, built on piles, of Pskovshchina, Novgorodchina and Estonia.

3. In talus slopes and strata of loess and loess-like loam, many millions of scattered mammalian bones were buried. Early humans contributed to their concentration in certain places. They constructed dwellings from mammoth bones and were storing mammoth tusks 200 centuries ago. Bones from the loess of France, Romania, and the Ukraine are friable and rough from the surface, because of lying outdoors under variable water and oxygen conditions for such a long time. They are off-white

from the surface and in fracture, usually with dots and winding threads of manganese deposits along canalicules. The tusks and teeth of mammoths in loess defrosted a long time ago and in the talus deposits of Europe and Southern Siberia are of poor preservation. They are fractured and rotten, sometimes resembling marl or chalk. In the zone of permafrost, tusks and bones are of two types. From loess, the bones are pale yellow (rarely), usually orange, or orange-brown, and covered with a thin layer of asovskite—a recently described mineral from a group of iron oxides. From lacustrine-swamp deposits the bones are covered in blue by the mineral vivianite. On the sea shelf, tusks of both types acquire a brown coloring because of rewashing by saltwater.

4. The mass death of animals and their burial in accumulations of oxidized and thickened oil is a sad reality, indebted to the wastefulness of humans. In the Pleistocene this death from being mired in tar was very local. Bones from bitumens and ozokerites are dark brown, colored through and through with oil. From the ozokerite of Staruna (Ukraine) are corpses of rhinoceros and mammoths. From Tatariya and the Caucasus (Binagada) there is a famous collection of beetles, bones of birds and various animals. The enamel of teeth is dark blue (blue-gray) because of oil seeping underneath the enamel. Collagen was preserved almost in its entirety, but the odor was masked by that of the oil.

5. The accumulation of bones in caves—the result of the drift of corpses by water currents and of the active dwelling in the caves by animals and humans. In the south of the Ukraine are rockshelters of Pliocene age, filled with bones of Tertiary mammals and birds. We know of caves in Karpat, Krym, the Caucasus, the Urals, Tyan-Shan and the area near the Baltic Sea. Bones from caves of arid regions are pale yellow, sometimes with an encrustation of lime, with few dendrites. Their edges slightly "jingle," like broken glazed pottery. Collagen is poorly preserved, leached, and therefore not detected by smell. In humid regions bones are heavy, have abundant dendrites, well preserved collagen and the smell of damp bone.

Excrements are characteristic of particular caves and grottos, most often belonging to hyenas and wolves. The large quantity of lime in them from eaten bones usually causes good preservation.

6. In human settlements of historical times, from lone carnivores to fortresses and large cities, osteological remains were accumulated in random middens and in organized dumping grounds, while being covered for centuries by sediments. The osteological matter in organized dumping grounds, such as pits or cellars, has changed little over a few centuries or even millennia in the arid/cold climate of the tundra, Sredizemnomore, Middle Asia, the taiga zone and in humid regions.

The bone middens of the Middle Age fortress Sarkel on the Don, where two hundred thousand bones of domestic and wild animals were examined, the multi-meter strata of shell dumping grounds of ancient Greek and Roman colonies on the shores of the Black Sea and other deposits have left a strong impression on us. All this gives abundant food for taphonomic thought.

SECONDARY DAMAGE OF THE BONES

1. Mechanical abrasion of the bones occurs most often in the wave-cut zone of lake deposits, specially in the layer of rubble and rock debris. "Distant movement" in the water current is not required for this. Bones in moraines and in deserts often are strongly worn because of wind erosion.

2. Dissolution of the bone surface occurs in streams with the formation of "pockmarked" relief.

3. Exfoliation of the surface and formation of cracks in the bone is the result of lengthy

exposure to the sun, rain, and frost before being covered with sediments.

4. Biogenic and anthropogenic damage to bones occurs of several types:

Chemical dissolution of the surface of bones with the formation of winding lines occurs in the soil layer by plant roots.

Tooth marks. Well distinguished are the tooth marks of bones by carnivores—wolves, hyenas, wolverines—and ungulates, especially deer. Ancient rodents, which loved to gnaw through the skulls of people, deliver many pleasures to archaeologists and medics. Such operations until now were considered among the top of surgical art in the Neolithic, being exhibited in museums to an astonished public.

Mining of bones in sediments and on the bottom of water bodies is conducted by the larvae of insects and mollusks.

Grooves, notches, polishing of bones, and the manufacture of instruments from them by ancient masters can be aged to some extent by the existence of patina. This should always be considered by archaeologists.

DISEASE OF BONES

While dying out the mammoths were ill. Their tusks and molars acquired deformed shapes. Such formations are famous from the Novosibirsk islands.

Also sick with arthritis were cave bears and cave hyenas. The microclimate of caves became more damp and drafty. On the metapodia and phalanges of bears, it is simply pitiful to see. Not uncommon as well are cases of caries of their teeth. Hyenas, in addition, consumed an abundance of calcium, and their tendons are characterized by boney outgrowths, warts and spokes.

The frontal bones of several skulls of our Pleistocene bison are worn with holes, as if shot through by a laser. Actually these knotholes are the result of perforation of the bone by insect larvae or nematodes. Such openings are also on the skulls of cave lions and wolves.

CONCLUSION

I have reported a complex of visual features (qualities) in fossil bones from various types of Quaternary locations. Concrete examples were cited of the physical properties of bone matter, which were acquired by them in the process of burial in sediments, namely a change in specific weight, chemical composition, color, odor, feel, sound and physical shape. Elementary techniques are recommended for the best perception. Our method enables one to distinguish among bones of the early, middle, and late Pleistocene and Holocene.

Chapter 6

Taphonomy of a Late Pleistocene Carnivore Den, Dade County, Florida

STEVEN D. EMSLIE* and GARY S. MORGAN†

*Florida Museum of Natural History, University of Florida, Gainesville, Florida 32611
†New Mexico Museum of Natural History, 1801 Mountain Road NW, Albuquerque, NM 87104

INTRODUCTION

The Cutler Hammock Local Fauna is a late Pleistocene (Rancholabrean) fossil deposit located 13 km southwest of Miami on the Atlantic Coastal Ridge, Dade County, Florida (Fig. 1), and approximately 5 m above present sea level. The deposit, discovered in 1985 when amateur archaeologists found fossil bones on the surface, fills a large sinkhole/cave in the Miami Limestone. Shortly after its discovery, the Dade County Historic Preservation Division directed excavations at the site to recover artifacts and other material before they were lost or destroyed. The resulting salvage excavations revealed an abundance of fossil material and vertebrate paleontologists from the Florida Museum of Natural History (FLMNH) were alerted to the site and participated in some of the fieldwork. This work continued until December 1986 and produced a large collection of vertebrate fossils, human remains and archaeological artifacts dating to the late Pleistocene and early Holocene.

It was the discovery of human bone in association with fossils of extinct Pleistocene vertebrates, including dire wolf (*Canis dirus*) and other carnivores, that brought Dr. Paul S. Martin to the site in February 1986. His interest in this site provided the impetus to present for the first time a description and taphonomic interpretation of the fossil material from Cutler Hammock. Here, we provide a preliminary analysis of the vertebrate remains from this unique site. Specifically, we discuss the complex taphonomy of the fossil deposits that appear to have been formed through a combination of numerous abiotic and biotic processes. These processes include long-term use of the site by dire wolves and other extinct carnivores during the late Pleistocene, and by humans in the early Holocene. We also provide the first evidence of the diet of dire wolves.

SITE DESCRIPTION AND EXCAVATIONS

The sinkhole containing the fossil deposits is approximately 5 x 6 m in diameter at the surface, and at least 3 m deep (Fig. 2). It is

Late Quaternary Environments and Deep History: A Tribute to Paul S. Martin, edited by David W. Steadman and Jim I. Mead. The Mammoth Site of Hot Springs, South Dakota, Inc. Scientific Papers, Volume 3. Hot Springs, South Dakota, 1995.

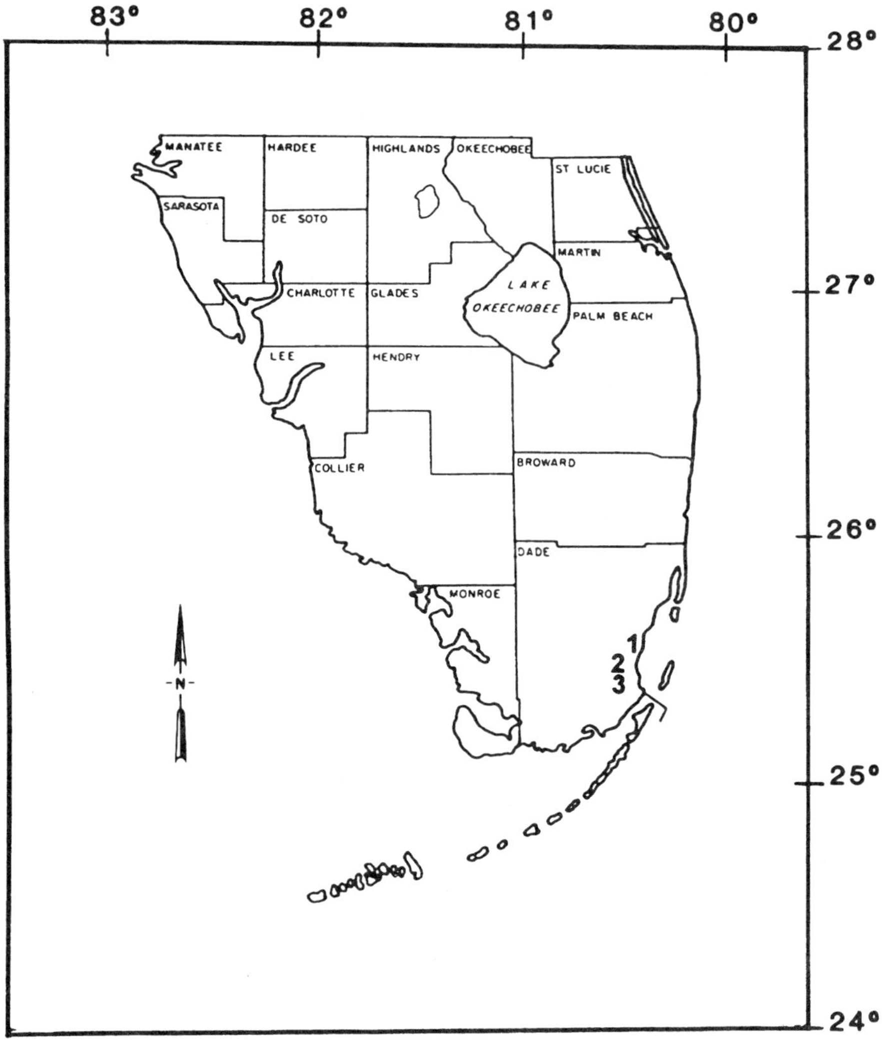

FIG. 1. Map of southern peninsular Florida and the Florida Keys showing the location of the fossil sites from Dade County discussed in the text. 1. Cutler Hammock, late Rancholabrean; 2. Monkey Jungle Hammock, late Rancholabrean and Monkey Jungle 2, Holocene; 3. Nichols Hammock, Holocene.

located within a tropical hardwood hammock on the Cutler Ridge, 0.3 km west of the Atlantic coast. This hammock is typical of other hardwood hammocks found in southern Florida that are dominated by tropical evergreen and semi-evergreen trees, most of which also occur in the West Indies and Neotropics, but not in northern Florida (Snyder *et al.*, 1990). Common trees in these hammocks include gumbo limbo (*Bursera simaruba*), pigeon plum (*Coccoloba diversifolia*), white stopper (*Eugenia axillaris*) and live oak (*Quercus virginiana*). Other species include strangler fig (*Ficus aurea*), West Indies cherry (*Prunus myrtifolia*), inkwood (*Exothea paniculata*), bitterbush (*Picramnia pentandra*) and Jamaica dogwood (*Piscidia piscipula*). These hammocks often occur as small islands on elevated areas near flooded prairies, marshes or coastline (Snyder *et al.*, 1990).

Since its formation in the late Pleistocene, the Cutler Hammock sinkhole has filled with sediments and limestone rock. Much of these sediments formed a cone or sediment dome of which the lower half is below the current water table and includes portions of indurated soil, rock and fossils. The upper half contains fewer fossils and was disturbed by pot hunters.

The site was excavated by placing a 22 m^2 grid system over the surface of the sinkhole (Fig. 2; Carr, 1987). Most of these squares were excavated to at least 2 m in depth. Probing and coring of the sediments indicated that an additional 3 m of deposit was present in some areas. At approximately 1.25 m depth, a layer of burned rock, bone and artifacts was discovered that included a hearth and at least one scattered human burial. Two charcoal samples were collected from this layer of which one yielded a radiocarbon date of 9670 ± 120 yr B.P. (Beta-16750; Carr, 1987).

All sediments excavated from the sinkhole were sifted through 1/4-inch and window screen, resulting in the recovery of thousands of bones of small mammals, reptiles, amphibians and birds. In addition, excavations produced thousands of bones of large mammals and birds. All fossil material is housed at the Florida Museum of Natural History, Gainesville. Archaeological artifacts and human bones are deposited at the Historical Museum of Southern Florida, Miami. At the conclusion of the project, the sinkhole was filled with sand covering the unexcavated portion of the site. The property on which the Cutler Site is located has been acquired by the Deering Estate Park thereby protecting it from future disturbances.

The age of the locality was determined by geology and mammalian biochronology. The Miami Limestone formed during the last interglacial period, or approximately 130,000 to 120,000 yr B.P. (Osmond *et al.*, 1965). Thus, deposition in this newly formed karst topography could not have begun until the late Pleistocene. Extinct vertebrates in the fauna are typical of other localities in Florida and North America that have been dated to the Rancholabrean Land Mammal Age and that became extinct at approximately 11,000 yr B.P. These species include *Canis dirus*, Florida cave bear (*Tremarctos floridanus*), sabertooth cat (*Smilodon fatalis*), horse (*Equus* sp.), Columbian mammoth (*Mammuthus columbi*)

FIG. 2. View of sinkhole at Cutler Hammock during excavations in 1985. The sinkhole was filled with sediments and limestone rock as shown here. Wood beams are part of the grid system used to excavate the site.

and American mastodont (*Mammut americanum*). Several species from Cutler Hammock are restricted in Florida to late Rancholabrean faunas and include bison (*Bison antiquus*), American lion (*Panthera atrox*), puma or Florida panther (*Puma concolor*), cotton rat (*Sigmodon hispidus*), pine vole (*Pitymys pinetorum*) and fox squirrel (*Sciurus niger*) (Morgan, 1991; Morgan and Hulbert, in press.)

THE VERTEBRATE FAUNA AND PALEOECOLOGY

Our preliminary list of the vertebrate fauna from Cutler Hammock includes over 100 species and represents all vertebrate classes (Table 1). We stress that this list is preliminary and will change with additional research on this extensive collection. In addition, most of these taxa remain unstudied and exact numbers of bones and minimum number of individuals (MNIs) have not been calculated. To date, only the bat fauna has been reported (Morgan, 1991).

The most abundant bones from the site are of amphibians and reptiles. The most common members of the herpetofauna are the spadefoot toad (*Scaphiopus holbrooki*) and colubrid snakes, especially the indigo snake (*Drymarchon corais*). The amphibian remains in particular suggest that water was available in the sinkhole at least for part of the year during formation of the deposits. However, the rarity of obligate freshwater species in the herpetofauna suggests that the water in or near the site probably was ephemeral. Shell fragments of terrestrial tortoises and turtles are common in the Cutler Hammock site, including a small extinct species of land tortoise (*Hesperotestudo incisa*), gopher tortoise (*Gopherus polyphemus*) and box turtle (*Terrapene carolina*), but freshwater turtles are rare. The absence of the giant land tortoise (*Hesperotestudo crassiscutata*) is notable, as this species is one of the most common large vertebrates in other Florida Rancholabrean faunas, especially in cave, sinkhole, and fissure deposits.

The avifauna is comprised of at least 23 extant and 7 extinct species that reflect a diversity of environments including lakes, marshes and coastal habitats, open grasslands, pinelands and hardwood hammocks. Of the extant taxa, one (*Gymnogyps californianus*) is found today only in the western U.S. while two others occur in Florida, but not in the region of the site. The Crested Caracara (*Polyborus plancus*) is restricted to open grasslands and palmetto regions near Lake Okeechobee approximately 100 km to the north of Cutler Hammock (Kale and Maehr, 1990). The Cooper's Hawk (*Accipiter cooperii*) breeds in forested areas only as far south as Lake Placid or approximately 150 km north of the site. In addition to the condor and caracara, the merlin (*Falco columbarius*) and Peregrine Falcon (*F. peregrinus*) occur primarily in open areas including grasslands and marshes (Kale and Maehr, 1990) and indicate the presence of open country near this site during the Late Pleistocene.

One extinct caracara (*Milvago readei*) is well-represented in this avifauna. This taxon is known from three other Rancholabrean sites in Florida: Arredondo (Brodkorb, 1959), Ichetucknee River (Campbell, 1980) and Reddick 1A (1 bone, a distal left tibiotarsus, UF 95900, personal observation). Its living counterparts are the Yellow-headed Caracara (*M. chimachima*), which inhabits grasslands and savanna in Central and South America, and the Chimango Caracara (M. *chimango*) found in open habitats and forest edges throughout South America. Another extinct species is the hawk-eagle (*Spizaetus grinnelli*) which in skeletal morphology is very similar to the living Ornate Hawk-eagle (*S. ornatus*) found in forested areas of South America. The fossil species is known from only one other locality in Florida, West Palm Beach, Palm Beach County

Table 1. Preliminary list of vertebrate taxa identified from the late Rancholabrean Cutler Hammock Local Fauna, Dade County, Florida. Systematic arrangement and species names follow Hulbert (1992). See text for data on relative abundance and number of individuals for selected taxa.

- Class Chondrichthyes
  - Order Lamniformes
    - Sand Shark (*Odontaspis taurus*)
- Class Osteichthyes
  - Order Lepisosteiformes
    - Garfish (*Lepisosteus* sp.)
  - Order Amiiformes
    - Bowfin (*Amia calva*)
  - Order Perciformes
    - Great Barracuda (*Sphyraena barracuda*)
    - Tuna (*Thunnus* sp.)
- Class Amphibia
  - Order Caudata
    - Siren (*Siren* sp.)
    - Amphiuma (*Amphiuma* sp.)
    - genus and species undet.
  - Order Anura
    - Eastern Spadefoot Toad (*Scaphiopus holbrooki*)
    - Robber Frog (cf. *Eleutherodactylus* sp.)
    - Southern Toad (*Bufo* cf. *B. terrestris*)
    - Treefrog (*Hyla* sp.)
    - Frog (*Rana* sp.)
- Class Reptilia
  - Order Testudines
    - Kinosternidae, genus and species undet.
    - Eastern Box Turtle (*Terrapene carolina*)
    - Pond Turtle (*Pseudemys* sp.)
    - Gopher Tortoise (*Gopherus polyphemus*)
    - †Land Tortoise (*Hesperotestudo incisa*)
  - Order Squamata
    - Green Anole (*Anolis carolinensis*)
    - Indigo Snake (*Drymarchon corais*)
    - Colubridae, genus and species undet. (5 spp.)
    - Eastern Diamondback Rattlesnake (*Crotalus adamanteus*)
- Class Aves
  - Order Gaviiformes
    - Common Loon (*Gavia immer*)
  - Order Podicipediformes
    - Pied-billed Grebe (*Podilymbus podiceps*)

Order Pelecaniformes
- †Anhinga (*Anhinga* sp.-large)

Order Ciconiiformes
- †Stork (*Ciconia maltha*)
- Black Vulture (*Coragyps atratus*)
- *California Condor (*Gymnogyps californianus*)
- †Teratorn (*Teratornis merriami*)

Order Anseriformes
- Mallard (*Anas platyrhynchos*)
- Ruddy Duck (*Oxyura jamaicensis*)

Order Accipitriformes
- Bald Eagle (*Haliaeetus leucocephalus*)
- #Cooper's Hawk (*Accipiter cooperii*)
- Red-shouldered Hawk (*Buteo lineatus*)
- Hawk (*Buteo* sp.)
- †Hawk-eagle (*Spizaetus grinnelli*)
- #Crested Caracara (*Polyborus plancus*)
- †Caracara (*Milvago readei*)
- Merlin (*Falco* cf. *F. columbarius*)
- Peregrine Falcon (*Falco* cf. *F. peregrinus*)

Order Galliformes
- Wild Turkey (*Meleagris gallopavo*)
- Northern Bobwhite (*Colinus virginianus*)

Order Gruiformes
- King Rail (*Rallus elegans*)
- American Coot (*Fulica americana*)

Order Charadriiformes
- Common Snipe (*Gallinago gallinago*)

Order Columbiformes
- Pigeon (*Columba* sp.)
- Mourning Dove (*Zenaida macroura*)
- Dove (*Zenaida* sp.)
- †Passenger Pigeon (cf. *Ectopistes migratorius*)

Order Strigiformes
- Barn Owl (*Tyto alba*)
- Eastern Screech-owl (*Otus asio*)
- Barred Owl (*Strix varia*)

Order Piciformes
- Woodpecker (*Dryocopus* sp.)

Order Passeriformes
- †Jay (cf. *Protocitta dixi*)
- Fish Crow (*Corvus* cf. *C. ossifragus*)
- genus and species undet.

Class Mammalia

Order Marsupialia

Opossum (*Didelphis virginiana*)

Order Xenarthra

†Beautiful Armadillo (*Dasypus bellus*)

†Northern Pampathere (*Holmesina septentrionalis*)

†Harlan's Ground Sloth (*Paramylodon harlani*)

Order Insectivora

Least Shrew (*Cryptotis parva*)

Eastern Mole (*Scalopus aquaticus*)

Order Chiroptera

#Southeastern Myotis (*Myotis austroriparius*)

#Big Brown Bat (*Eptesicus fuscus*)

Evening Bat (*Nycticeius humeralis*)

*Peter's Ghost-faced Bat (*Mormoops megalophylla*)

Order Carnivora

Gray Fox (*Urocyon cinereoargenteus*)

#Coyote (*Canis latrans*)

†Dire Wolf (*Canis dirus dirus*)

†Florida Cave Bear (*Tremarctos floridanus*)

Black Bear (*Ursus americanus*)

Raccoon (*Procyon lotor*)

River Otter (*Lutra canadensis*)

Spotted Skunk (*Spilogale putorius*)

Striped Skunk (*Mephitis mephitis*)

†Sabertooth Cat (*Smilodon fatalis*)

Bobcat (*Lynx rufus*)

*Jaguar (*Panthera onca*)

†American Lion (*Panthera atrox*)

Florida Panther (*Puma concolor*)

Order Rodentia

Eastern Gray Squirrel (*Sciurus carolinensis*)

Fox Squirrel (*Sciurus niger*)

Southern Flying Squirrel (*Glaucomys volans*)

#Southeastern Pocket Gopher (*Geomys pinetis*)

Hispid Cotton Rat (*Sigmodon hispidus*)

#Eastern Wood Rat (*Neotoma floridana*)

Florida Mouse (*Podomys floridanus*)

Oldfield Mouse (*Peromyscus polionotus*)

Cotton Mouse (*Peromyscus gossypinus*)

Rice Rat (*Oryzomys palustris*)

Round-tailed Muskrat (*Neofiber alleni*)

#Pine Vole (*Pitymys pinetorum*)

Order Lagomorpha

Eastern Cottontail (*Sylvilagus floridanus*)

Order Artiodactyla

†Long-nosed Peccary (*Mylohyus nasutus*)

†Flat-headed Peccary (*Platygonus compressus*)
†Large-headed Llama (*Hemiauchenia macrocephala*)
†Stout-legged Llama (*Palaeolama mirifica*)
White-tailed Deer (*Odocoileus virginianus*)
†Bison (*Bison antiquus*)
Order Perissodactyla
†Horse (*Equus* sp.-small)
†Horse (*Equus* sp.-large)
Order Proboscidea
†American Mastodont (*Mammut americanum*)
†Columbian Mammoth (*Mammuthus columbi*)

† extinct species
* extinct in Florida, but extant elsewhere in North America
# occurrence is outside present range of this species in Florida

(Becker, 1985). This species, with *M. readei*, add to a growing list of tropical to subtropical species of vertebrates that are known from the Rancholabrean of Florida.

Most of the birds from Cutler Hammock commonly inhabit forests and hammocks today. These species include *Accipiter cooperii*, Red-shouldered hawk (*Buteo lineatus*), Wild Turkey (*Meleagris gallopavo*), Northern Bobwhite (*Colinus virginianus*), pigeon (*Columba* sp.), Eastern Screech-Owl (*Otus asio*), Barred Owl (*Strix varia*) and woodpecker (*Dryocopus* sp.) Other species indicate the nearby presence of marshes, lakes or other wetlands and include Common Loon (*Gavia immer*), Pied-billed Grebe (*Podilymbus podiceps*), anhinga (*Anhinga* sp.), Mallard (*Anas platyrhynchos)*, Ruddy Duck (*Oxyura jamaicensis*), American Coot (*Fulica americana*) and King Rail (*Rallus elegans*). Some of these species may have been brought to the site by the raptors and scavengers that also are represented here. Hundreds of bird bones remain to be identified from this rich assemblage.

The mammalian fauna also is rich and is comprised of at least 31 extant and 16 extinct species. The extant taxa include two that presently do not occur in Florida (*Mormoops megalophylla* and *Panthera onca*), and six currently in Florida, but not in the region of the site (see Table 1). The pocket gopher (*Geomys pinetis*), pine vole (*Pitymys pinetorum*), coyote (*Canis latrans*) and two species of bats (*Eptesicus fuscus* and *Myotis austroriparius*) are now restricted to the northern two-thirds of the state. At least two of these species require specialized habitats that are no longer present in southern peninsular Florida on the Florida Keys. Pocket gophers prefer areas with deep sandy soils for burrowing and the southeastern bat *M. austroriparius* is an obligate cave-dweller for at least part of the year. Dry caves suitable for bat roosts are currently absent from the southern two-thirds of the Florida peninsula (Morgan, 1991).

The most abundant taxa of large herbivores are peccary (*Mylohyus nasutus*, MNI=75 adult and 24 juveniles) and *Equus* sp. (2 adults, 17 juveniles). The abundance of *M. nasutus* at Cutler Hammock is particularly notable as this species previously was known only by remains of one or a few individuals from each of at least 35 Irvingtonian and Rancholabrean localities in North America (Kurtén and Anderson,

1980). It is thought to have been a solitary animal that inhabited forest edges, but did not use caves for shelter. Of the carnivores, *Canis dirus* is well-represented (see below) followed in abundance by *Tremarctos floridanus* (3 adults, 6 young adults), *Canis latrans* (5 adults), *Panthera onca* (4 adults) and *Lynx rufus* (3 adults). The samples of *C. dirus, M. nasutus* and *T. floridanus* are the largest (based on MNI's) of these species from eastern North America. As with the avifauna, hundreds of mammalian fossils remain to be identified.

The reason for the abundance of vertebrate taxa in this fauna may relate to the geography and paleoecology of southern Florida during the late Pleistocene. Previous studies have shown that sea level was 100 m lower during the glacial maximum at about 15,000 yr B.P. (Bloom, 1983). This drop in sea level exposed large areas of the shallow continental shelf, especially the Florida Keys and off the Gulf coast. Peninsular Florida was probably twice as large as today (Webb and Wilkins, 1984). Cutler Hammock would have been situated on a high ridge in the late Pleistocene with the Atlantic coast farther to the east. Moreover, the water table may have been 26-31 m lower than today (Watts and Hansen, 1988). Since the Miami Limestone underlies this area, a well-developed karst topography with extensive cave systems probably existed here, explaining the high diversity of bats in Rancholabrean deposits (Morgan, 1991).

The vertebrate fauna from Cutler Hammock provides additional paleoecological information on this region. The abundance of forest-dwelling species suggests that a hardwood hammock and/or pinelands existed near the sinkhole, similar to that of today. In addition, regions west and south of the site probably consisted of open grassland/savanna, pineland and lowland marshes and wetlands where the Everglades currently occur. Extensive grasslands would have supported herds of horse, bison, mammoth and other herbivores that are well-represented in the site.

One other late Pleistocene site in southern Florida corroborates the paleoenvironmental conditions shown by the Cutler fauna. At Monkey Jungle Hammock, 12 km southwest of Cutler (Fig. 1), fossils from a sinkhole located in a hardwood hammock represent 43 species of mammals including *Canis dirus, Tremarctos floridanus, Panthera atrox, Puma concolor* and *Pitmys pinetorum* (Morgan, 1991). This site and Cutler Hammock are the southernmost Rancholabrean sites in the continental U.S. Their latitudinal position (between 25° and 26° N) is reflected by the subtropical species in the faunas including the birds discussed above, *Panthera onca* and four species of bats (*Mormoops megalophylla, Pteronotus* cf. *P. pristinus, Eumops glaucinus,* and *Tadarida brasiliensis*; Morgan, 1991). Two other fossil localities in southern Florida appear to be Holocene in age. Monkey Jungle 2 and Nichols Hammock, Dade County (Fig. 1), are sinkhole deposits similar to Cutler Hammock but have produced no remains of extinct species of Rancholabrean age (Morgan, 1991). Hirschfeld (1968) also reported *Tadarida brasiliensis* from Nichols Hammock.

TAPHONOMY

Investigations at the Cutler Hammock indicate that this site formed by numerous processes and has a complicated taphonomy. Evidence for these processes is provided by the diversity, relative abundance and age structure of selected mammalian species, preservation and breakage patterns of bones from large mammals, and the human occupation of the site in the early Holocene. These abiotic and biotic processes are difficult to separate in terms of their relative influence on the formation of the fossil deposit. Here, we discuss the processes that were probably most important for bone accumulation at this site. In particular, we investigate the role of carnivores at this site and the possibility that the sinkhole functioned as a den, lair or feeding area by one or more species in the late Pleistocene.

It is probable that for at least part of its developmental history, the sinkhole was a natural trap for small vertebrates including toads and snakes. Thousands of bones of these taxa were recovered from the deposits where these animals may have become trapped in the bottom of the sinkhole where standing water was located. Frogs are known to live in or use caves (Lee, 1969), and thousands of young spadefoot toads (*Scaphiopus holbrooki*) have been observed trapped in Warren Cave, Alachua County, Florida (R. Franz, personal communication). A similar situation may have occurred at Cutler Hammock.

Equally possible is that raptors preyed upon these species and used the sinkhole, or trees above it, for roost or nest sites. Most raptors produce pellets of bone, hair, feathers and other indigestible parts of their prey which are regurgitated at the roost/nest. These pellet accumulations can be substantial and add hundreds of bones to a deposit. The diversity of raptorial and scavenging birds in the avifauna at Cutler Hammock suggests that these birds used this site and may have been responsible for at least some of the microfauna deposited there. In particular, hawks (*Buteo* spp.), falcons (*Falco* spp.), bald eagle (*Haliaeetus leucocephalus*), owls (*Otus asio, Strix varia, Tyto alba*), vultures (*Coragyps atratus, Gymnogyps californianus*) and caracara (*Polyborus plancus*) occur in the site and are known to form pellets and accumulate bone at their roosts and nests (Moon, 1940; Dodson and Wexlar, 1979; Brain, 1981; Emslie, 1987, 1988; Andrews, 1990).

Because birds can forage over long distances, Andrews (1990) pointed out that many small vertebrates in fossil sites may not represent local environments. Long-distance transport of bones by birds has been documented for fossil California condor (Emslie, 1988) and Western Gulls (Emslie and Messenger, 1991). Moreover, long-distance transport of prey by leopards also has been documented (Brain, 1981). Caution should precede paleoecological interpretations based on one or more "unusual" species in a fauna, especially when bone-collecting raptors are present in the assemblage.

Mammalian predators add bone to deposits in caves and sinkholes where they may den or seek temporary shelter. Specific carnivores have been documented depositing remains of small vertebrates in scat (Mellett, 1974; Mead and Van Devender, 1981) or simply by carrying food to young at den sites (Brain, 1981). It is this latter means of transport that may have caused most of the large mammal remains to be deposited in Cutler Hammock, especially with the diversity of large carnivores also found there. These carnivores include three canids (one extinct), five felids (three extinct), and two bears (one extinct, Table 1). This high diversity of large carnivores at one site is unusual; Cutler Hammock is one of the few sites in North America that documents the co-occurrence of four large felids, including *Puma concolor, Panthera atrox, Panthera onca* and *Smilodon fatalis*. Here, we concentrate our discussion on the use of this site by large carnivores and their probable role in bone accumulation at Cutler Hammock.

The role of carnivores in fossil deposits

Determining the influence of carnivores on fossil accumulations is facilitated by numerous studies that have been completed at modern dens and rendezvous sites of the wolf (*Canis lupus*), hyaena (*Crocuta, Hyaena*) and leopard (*Panthera pardus*). Research by Haynes (1980a, 1982), Brain (1980, 1981), Binford (1981), Hill (1989), Skinner and van Aarde (1991) and Lam (1992) in addition to other studies cited by these authors have been valuable analogs for comparing fossil accumulations produced by carnivores.

According to Binford (1981), a carnivore den will have bones with the following characteristics: (1) the number of bone splinters will be inversely correlated with the number of articular ends as most of these ends are destroyed by gnawing, (2) bones will be marked by pitting, furrowing, channeling and other damage caused by gnawing that does not occur at kill sites (see also Haynes, 1980a), (3) only large splinters, > 4 cm long, will show characteristics of gnawing while small splinters may show evidence of acid etching from stomach acids, (4) many breaks will occur at protrusions of thin ridges on bone resulting in long splinters with triangular cross-sections, such as the tibial crest, and (5) gnaw patterns may vary with specific skeletal elements. Similar patterns may be expected in a fossil site except that, in regard to point 3, large splinters may be further broken by additional gnawing, trampling or diagenetic processes thereby producing small splinters with gnaw marks as well. Since Binford was dealing with modern dens, these additional processes on the bone were not considered.

Identification of these patterns in a fossil deposit will reflect the role of carnivores in producing that deposit. In addition, Klein (1980) suggests that carnivore dens will have more remains of carnivores, both in diversity and MNI's, than other kinds of sites including human. However, natural traps may have high carnivore diversity as well (Martin and Gilbert, 1978; White *et al*., 1984) and were not considered by Klein. This bias toward carnivores in traps is probably caused by their attraction to dead and dying prey that also are in the trap. The fossil accumulation at Rancho La Brea, heavily biased towards carnivores (Stock, 1942; Marcus, 1960), provides another example of this pattern.

We believe the carnivore assemblage at Cutler Hammock is not due to a natural trap for two reasons. First, hundreds of bones and bone fragments of other animals (e.g., *Equus, Mylohyus, Odocoileus*) have breakage patterns and gnaw marks similar to those described by Binford (1981) and Haynes (1983) that indicate use of the site for feeding by carnivores over a long period. Second, the age structure of predators and prey can be determined by examining tooth wear and eruption stages. These studies are described below and indicate that the site was used over a long-period by carnivores and could not have been a trap for these species. If Cutler Hammock was a natural trap, it must have acted as such only for small vertebrates as discussed above.

THE DIRE WOLF AT CUTLER HAMMOCK

Dire wolves are known from over 80 Rancholabrean localities in North and South America (Kurtén and Anderson, 1980). They are most abundant from the tar pits at Rancho La Brea where an estimated 1646 individuals are represented (Marcus, 1960). In addition, they are common in cave deposits. A large sample that includes 27 cranial and 30 mandibular fragments is known from San Josecito Cave, México (Nowak, 1979). In Florida, small samples are known from at least 19 other localities (Nowak, 1979; Webb, 1974; Morgan, 1991). The abundance of this species throughout North America allowed Kurtén (1984) to differentiate two distinct subspecies (*Canis d. dirus* and *C. d. guildayi*) based on size and limb proportions. In general, he found that the western subspecies (*C. d. guildayi*) was smaller, with proportionally shorter limbs and shorter but broader teeth compared to the eastern form (*C. d. dirus*). He attributed this difference to the adaptations of the eastern form to open plains.

The large sample of dire wolf at Cutler Hammock includes few complete limb bones, but numerous teeth are present and dental measurements allow comparison to *C. d. dirus* and

C. d. guildayi. Kurtén (1984) found significant differences in the length of the P^2 and width of P_2 and P_4 for late Wisconsinan populations of these subspecies. Measurements of width on eight left P_4's from Cutler Hammock ranged from 8.9-10.0 mm (mean ± s.d.=9.44 ± 0.35 mm) and are comparable to Kurtén's sample for *C. d. dirus* (9.61 ± 0.18). Based on these measurements, we refer all *C. dirus* from Cutler Hammock to *C. d. dirus*.

In modern canids, body size and dental measurements are larger in males with some overlap between both sexes (Nowak, 1979). Dental measurements from Cutler Hammock reflect sexual differences in size. Measurements of length and greatest width of the M_1, also used to calculate MNIs by age classes (see below), are plotted in Figure 3. The size distribution indicates a gradient from small to large with no clear distinction by sex. Teeth with measurements at the small or large end of the distribution probably represent female and male individuals, respectively. Figure 3 also indicates that there is no bias in the sample by sex and age class; males and females of young and aged adults appear to be represented in the sample. One exception is with young adults (Age Class 2) for which only small teeth are present. These small measurements, however, probably are due to the thin layers of cementum and dentine in teeth of young animals that thicken with age in canids (Linhart and Knowlton, 1967) rather than sexual differences in mortality.

We estimated individual age and MNIs of dire wolves at time of death by milk dentition and amount of tooth wear on the permanent M_1. Although tooth wear is not a reliable method for aging canids because of geographic variation in diet (Linhart and Knowlton, 1967), we believe it is appropriate for this single assemblage. In addition, our age classifications are defined on general rather than specific categories of tooth wear and do not attempt to re-

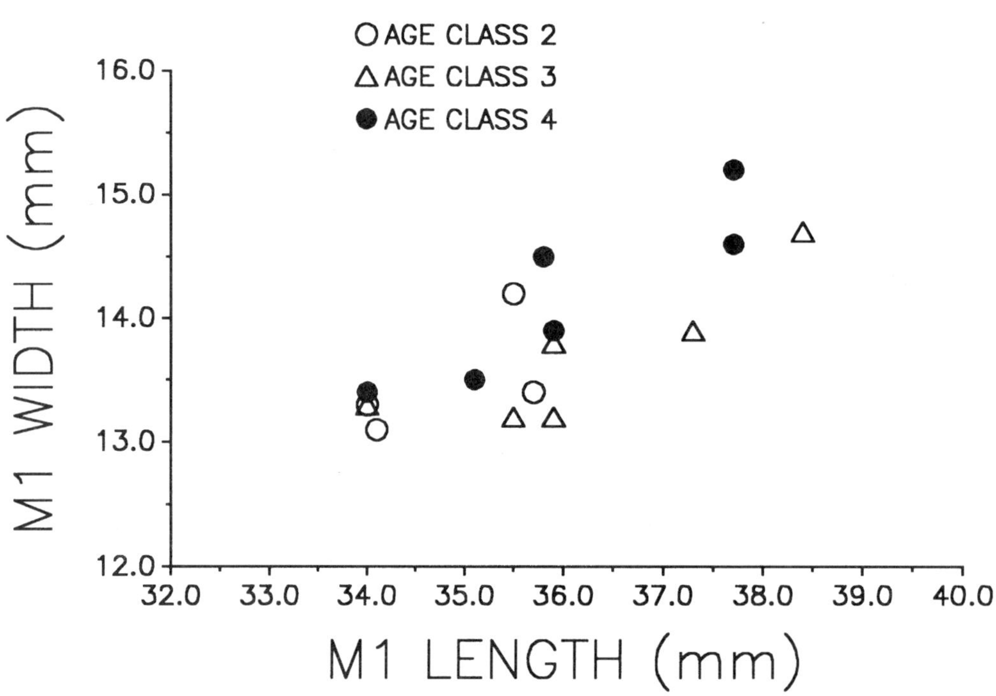

FIG. 3. Size distribution of adult individuals of *Canis dirus dirus* at Cutler Hammock based on length and greatest width of the M_1.

flect exact age in years. We use five categories in our analysis: (0) pre-weaned or milk dentition; (1) juveniles with permanent teeth erupting and unworn, (2) young adults with fully erupted permanent teeth showing little or no wear, (3) adults with moderate wear and (4) aged adults with extensive wear (Fig. 4-5). We believe these categories accurately reflect the general age structure of the fossil population.

At least 42 individuals of *Canis dirus* are represented at Cutler Hammock, though 13 of these are from milk dentition (DP^4) and may

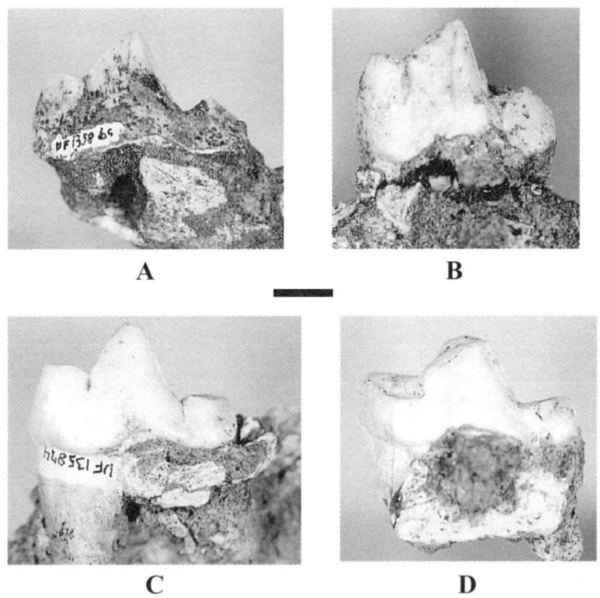

FIG. 4. Wear patterns on the left M_1 of *Canis dirus dirus* at Cutler Hammock that were used to estimate age classes of individuals. (A) Age Class 1 (UF 135869), juvenile with newly erupted permanent teeth with no wear; (B) Age Class 2 (UF 135862), young adult with little wear; (C) Age Class 3 (UF 135874), adult with moderate wear; (D) Age Class 4 (UF 135881), aged adult with heavy wear.

FIG. 5. Complete left mandible (UF 135887), lateral view, of *Canis dirus dirus* from Cutler Hammock. Scale 0.5 x; bar = 2 cm.

be duplicated in the MNIs based on permanent teeth. The age structure has a significant bias towards very young (Age Classes 0-1; MNI=19 or 45% of total) individuals in the fossil population (X^2, $p<0.05$; Fig. 6). This structure, which also includes at least 8 very old individuals (Age Class 4, 19%), is characteristic of an attritional assemblage that reflects true death rates in a population that usually are higher in the very young or old age classes (Shipman, 1981). These data also indicate that the site was used by these wolves over a long period and was not a natural trap.

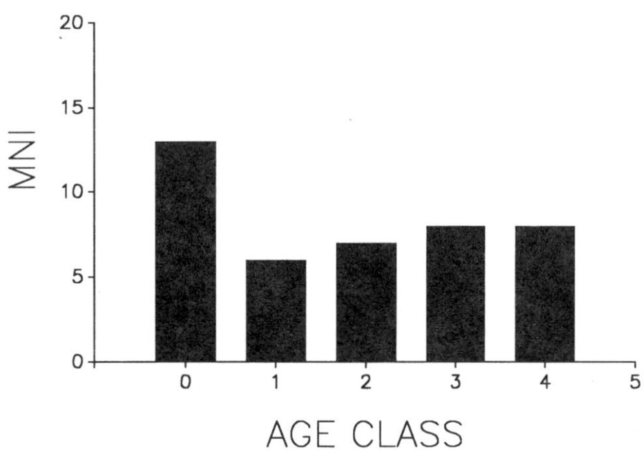

FIG. 6. The number of individuals of *Canis dirus dirus* identified in each age class from Cutler Hammock (see text for definition).

This age structure also may reflect how the site was used by this species. At modern wolf (*Canis lupus)* dens, the young are taken from the den to other locations within 8-10 weeks after birth, before milk dentition is replaced (Mech, 1970). These locations, called rendezvous sites, are activity areas near water or kill sites where the pups are fed, lose their milk teeth and are reared to independence. These sites may be used year after year by the pack whereas den sites, especially those located in forests, may be used only once (Haynes, 1980a). The 13 individuals of dire wolf that are represented by milk dentition at Cutler Hammock suggest that this locality was a rendezvous rather than a den site if these wolves practiced similar behavior as *C. lupus*. Thus, Cutler Hammock may have been an important feeding locale where prey also may have been abundant for dire wolves and other carnivores.

Evidence for feeding by dire wolves at Cutler Hammock is shown by the abundance of gnaw-damaged bone from a diversity of large herbivores. This sample includes hundreds of teeth and bones of *Mylohyus nasutus* and two species of *Equus*. The unusual abundance of *M. nasutus* (74 adults and 25 juveniles also based on teeth), as well as breakage patterns and marks on its bones (see below), suggests that this species may have been prey brought to the site by dire wolves. We estimate from a preliminary examination of the *Equus* teeth (deciduous and permanent) that at least 19 individuals of two species are represented of which 17 (90%) were very young or pre-weaned at time of death. Deer (*Odocoileus virginianus*) appears to be the third most abundant herbivore, followed by bison (*Bison antiquus*) and the camels (*Palaeolama mirifica* and *Hemiauchenia macrocephala*). Teeth of these last two species include deciduous and unworn permanent molars and premolars. One individual of mammoth (*Mammuthus columbi*) also is represented by two deciduous teeth. All these young animals would have been most susceptible to predation by dire wolves and other carnivores. This diversity of prey species represented at Cutler Hammock also is characteristic of a rendezvous rather than a kill or single-use den site (Haynes, 1980a).

A preliminary examination of breakage patterns and gnaw marks on the bones of these prey species appear to fit the characteristics for a wolf den or rendezvous site as described by Binford (1981) and Haynes (1980a, 1983). Distal ends of humeri of *Mylohyus* and *Odocoileus* (Fig. 7) display these patterns and can be compared to similar patterns shown by

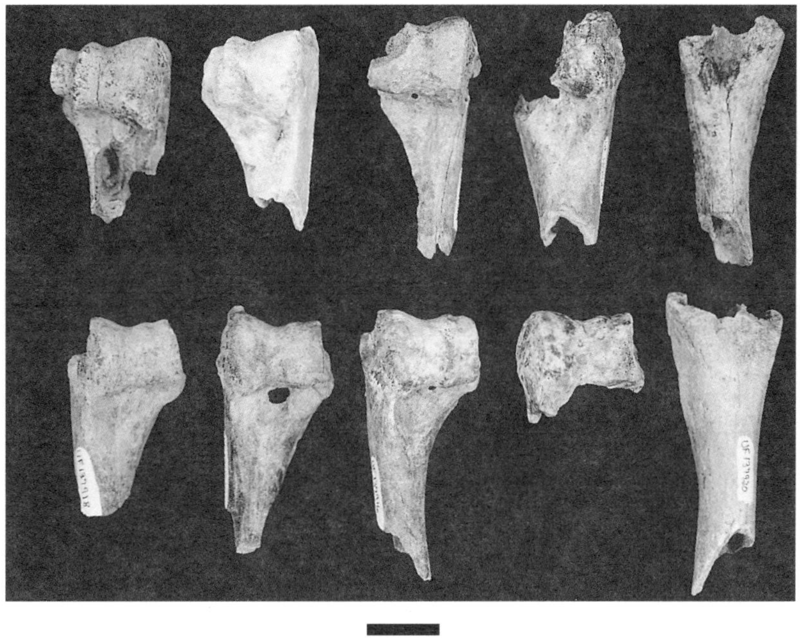

FIG. 7. Breakage pattern of distal humeri of artiodactyls from Cutler Hammock. Top row, left to right, UF 135900, 135896, 135895 (*Odocoileus virginianus*), 137921 and 137923 (Artiodactyla, genus and species indet.). Bottom row, left to right, UF 137918, 137917, 137916, 137915 (*Mylohyus nasutus*) and 137920 (Artiodactyla, genus and species indet.). Scale 0.5 x; bar = 2 cm.

hyaenas (Hill, 1989, his figs. 2 and 3) and wolves (Binford, 1981, his figs. 3.42 and 3.44). Moreover, the large number of unidentifiable bone splinters of large mammals from Cutler Hammock greatly exceeds that of articular ends. Marks on some of these ends include punctures and furrows (Fig. 8) that are most characteristic of damage caused by wolves, but not by bears or felids (Haynes, 1980b, 1983). These data suggest that dire wolves were most responsible for the breakage patterns and damage on bones of large herbivores at Cutler Hammock. Additional study of this fauna is needed to quantify these patterns in comparison to living species of wolves and other carnivores.

CONTRIBUTIONS OF OTHER CARNIVORES

The second most abundant carnivore at Cutler Hammock is *Tremarctos floridanus*. At least nine individuals are represented by comparing size and wear patterns of the M_2. Six of these individuals show little or no wear on their teeth and were relatively young in age at time of death; no milk dentition was recovered. This extinct bear and other tremarctine species from North America are represented today by

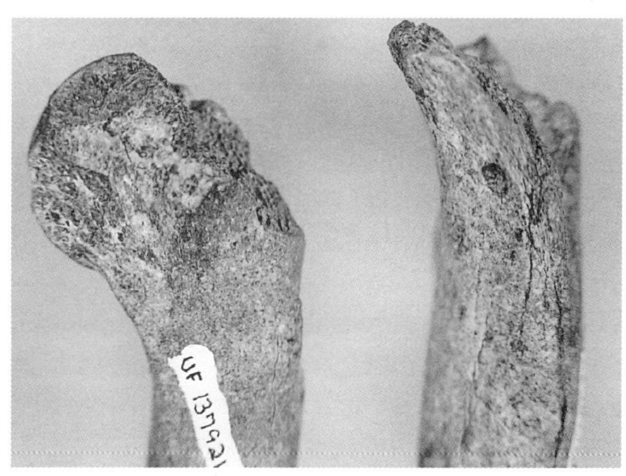

FIG. 8. Distal humeri of artiodactyls from Cutler Hammock showing damage caused by carnivore gnawing. Left, UF 137921, medial view; right, UF 137923, lateral view. Scale 1 x; bar = 1 cm.

a single species, the spectacled bear (*Tremarctos ornatus*) of South America. Using this latter species as an analog, Kurtén (1966) hypothesized that *T. floridanus* may have been herbivorous. This conclusion is based on similar morphology in dentition and cranial features that relate to diet in fossil and living species.

Tremarctos floridanus apparently used caves for denning or resting sites as its remains commonly occur in these locales. In Florida, Devil's Den has produced partial skeletons of at least five individuals (Kurtén, 1966). Although probably herbivorous, this bear may have opportunistically scavenged or preyed on other animals and may be responsible for some of the bones deposited in the site. Bears do leave characteristic gnaw marks on bone (Haynes, 1983), but we have not observed any bone from Cutler Hammock with unequivocal evidence of gnawing by bears. A detailed study of all bones of potential prey is needed to establish the role of bears in this fossil deposit.

Another common carnivore in the fauna is *Canis latrans*. At least five adult individuals are represented by the left P^4. This species may have used the site for a den, but no milk teeth have been identified. The coyote is known to take deer, elk and livestock and will scavenge from carcasses of larger animals (Bekoff, 1977). It may have added to the accumulation of bones of smaller prey at the site, including amphibians, reptiles, ground-dwelling birds, rodents and rabbits, as these species form a large part of its modern diet (Bekoff, 1977). Microfauna may also have been introduced to the site from its scat (Mellett, 1974). At Cutler Hammock, it may have preyed on peccary, deer and young horses whose remains are abundant in the site.

The fourth most abundant carnivore is *Panthera onca*, with at least four adult individuals. This species may have used the sinkhole as a lair or feeding area. Jaguars in South America today prey on peccaries as well as fish, turtles, small mammals and livestock (Nowak and Paradiso, 1983; Macdonald, 1984). It may have added remains of peccaries, deer and horse to the deposits.

At least three individuals of *Lynx rufus* are represented at Cutler Hammock. Bobcats are widespread in North America today where they prey primarily on small rodents, rabbits and ground birds (Macdonald, 1984). It potentially added remains of rodents, rabbits, quail, turkey and other small mammals to the deposits at Cutler Hammock.

The remaining large carnivores at Cutler Hammock are represented by only one individual each: *Ursus americanus, Puma concolor, Panthera atrox* and *Smilodon fatalis*. Each of these may have preyed or scavenged carcasses of larger mammals including horse, bison, sloth and elephants. Their poor representation in the fauna may indicate only occasional use of the sinkhole and a relatively small role in the accumulation of fossil material.

HUMAN BONES

Archaeological evidence at Cutler Hammock indicates human occupation in the early Holocene. This evidence includes a layer of burned bone, artifacts and human bones found approximately 1 m below the surface. The human bones, examined by G. Armelagos, are from three adults and two children (Carr, 1987). Most of the bones are fragmentary and those of one of the children are burned. A radiocarbon date on charcoal from this layer provided an age of 9670 ± 120 yr B.P. Artifacts, including points, knives and scrapers, are similar to the Dalton culture (9300 to 10,500 yr B.P.) in Florida and correspond in age with the ^{14}C date. Moreover, the source for the stone used in manufacturing these tools has been identified in north-central Florida over 300 km from Cutler Hammock (Carr, 1987).

Evidence for earlier human presence at the site is equivocal. Although human bones were found *in situ* in the lower deposits, 130-150 cm from the surface, and in direct stratigraphic association with bones of dire wolf and other extinct vertebrates, it is possible that the human bones were mixed into the lower deposits by the digging activities of land crabs, rodents, or other animals including humans. During excavations of the site in 1985-1986, land crabs were observed digging holes in the bottom of the sinkhole and have been present in this region probably since the early Holocene when modern coastlines were established. Rodent and carnivore use of the site also may have mixed bone from higher to lower stratigraphic positions. However, the human bones are fragmentary and exhibit some gnaw damage by carnivores. The possibility that dire wolves preyed on or scavenged carcasses of humans in the late Pleistocene cannot be ruled out.

A portion of a shaft of a distal humerus of human was submitted to the University of Arizona Radiocarbon Laboratory for analysis. A corrected date of 3040 ± 70 yr B.P. (AA 1723) was obtained on this bone which had a nitrogen content of 0.042%. According to Stafford *et al.* (1991), bone with nitrogen contents of less than 0.1-0.2% are likely to produce dates that are hundreds to thousands of years too young due to adsorption of exogenous organic matter. Given the stratigraphic position of this bone and the age of artifacts in the site, we believe this radiocarbon date is much younger than the true age of the bone. Unfortunately, the high amount of leaching of amino acids that has occurred to bone in the acidic soils of the Cutler Hammock precludes any attempts to obtain accurate dates on the bones of humans or other vertebrates with current technology.

CONCLUSIONS

The Cutler Hammock Local Fauna is an important late Rancholabrean site that has produced remains of over 100 vetebrate species. These species include five mammals and two birds with subtropical to tropical affinities that reflect the latitudinal position of this site. The fauna as a whole suggests a forested environment surrounded by open grassland and lowland marshes and wetlands. A high diversity and abundance of carnivores in the deposits indicate use of the sinkhole as a den or rendezvous site, especially by dire wolves. This use probably accounts for the bulk of the large mammal remains that are preserved here which, for the first time, provide evidence for the diet of dire wolves in the late Pleistocene of North America. Peccary, young horse and deer appear to have been its primary prey.

The role of other carnivores in the accumulation of bone at Cutler Hammock is unknown until more detailed studies on the fauna can be completed. The high diversity of ursids, canids, and felids in the deposits suggests that many carnivores may have been important in the taphonomy of the site. Raptorial and scavenging birds, also diverse and abundant in the fossil assemblage, probably added many bones as well, especially of smaller animals.

Other sources of bone accumulation may have been from the sinkhole forming a natural trap for small animals, at least for a part of its depositional history. The thousands of bones of amphibians and reptiles may have been deposited in the site in this manner. Some species probably used the sinkhole/cave for habitation (e.g., bats) as well. Human use of the site in the early Holocene probably added bones to the deposits as evidenced by burned bone of rodents, rabbits, deer and other species.

It is apparent from this preliminary study that the Cutler Hammock has an extremely complicated taphonomy. Here, we have documented the role that dire wolves and other carnivores may have had on this assemblage. Ongoing investigations of the vertebrate fauna should provide more detailed information on the taphonomy and paleoecology of this fossil deposit and the relative role of abiotic and biotic processes in its formation.

REFERENCES

Andrews, P. (1990). "Owls, Caves and Fossils." University of Chicago Press, Chicago.

Becker, J. (1985). A late Pleistocene (Wisconsinan) avifauna from West Palm Beach, Florida. *Bulletin of the British Ornithologist's Club* **105**, 37-40.

Bekoff, M. (1977). *Canis latrans. Mammalian Species* **79**, 1-9.

Binford, L. (1981). "Bones: Ancient Men and Modern Myths." Academic Press, New York.

Bloom, A.L. (1983). Sea level and coastal morphology of the United States through the late Wisconsinan glacial maximum. *In* "Quaternary Environments of the United States, Vol. 1. The Late Pleistocene" (S.C. Porter, Ed.), pp. 215-229. University of Minnesota Press, Minneapolis.

Brain, C.K. (1980). Some criteria for the recognition of bone-collecting agencies in African caves. *In* "Fossils in the Making" (A.K. Behrensmeyer and A.P. Hill, Eds.), pp. 108-130. University of Chicago Press, Chicago.

Brain, C.K. (1981). "The Hunters or the Hunted?" University of Chicago Press, Chicago.

Brodkorb, P. (1959). The Pleistocene avifauna of Arredondo, Florida. *Bulletin Florida State Museum, Biological Sciences* **4**, 269-291.

Campbell, K. (1980). A review of the Rancholabrean avifauna of the Itchtucknee River, Florida. *Natural History Museum of Los Angeles County, Contributions in Science* **330**, 119-129.

Carr, R.S. (1987). Early man in south Florida. *Archaeology*, Nov./Dec., 62-63.

Dodson, P., and Wexlar, D. (1979). Taphonomic investigations of owl pellets. *Paleobiology* **5**, 275-284.

Emslie, S.D. (1987). Age and diet of fossil California condors in Grand Canyon, Arizona. *Science* **237**, 768-770.

Emslie, S.D. (1988). Vertebrate paleontology and taphonomy of caves in Grand Canyon, Arizona. *National Geographic Research* **4**, 128-142.

Emslie, S.D., and Messenger, S.L. (1991). Pellet and bone accumulation at a colony of Western Gulls (*Larus occidentalis*). *Journal of Vertebrate Paleontology* **11**, 133-136.

Haynes, G. (1980a). Prey bones and predators: potential ecologic information from analysis of bone sites. *Ossa* **7**, 75-97.

Haynes, G. (1980b). Evidence of carnivore gnawing on Pleistocene and Recent mammalian bones. *Paleobiology* **6**, 341-351.

Haynes, G. (1982). Utilization and skeletal disturbances of North American prey carcasses. *Arctic* **35**, 266-281.

Haynes, G. (1983). A guide for differentiating mammalian carnivore taxa responsible for gnaw damage to herbivore limb bones. *Paleobiology* **9**, 164-172.

Hill, A. (1989). Bone modification by modern Spotted Hyaenas. *In* "Bone Modification" (R. Bonnichsen and M. Sorg, Eds.), pp. 169-178. Center for the Study of the First Americans, University of Maine, Orono.

Hirschfeld, S.E. (1968). Vertebrate fauna of Nichol's Hammock, a natural trap. *Quarterly Journal Florida Academy Science* **31**, 177-189.

Hulbert, R.C., Jr. (1992). A checklist of the fossil vertebrates of Florida. *Florida Museum of Natural History, Papers in Florida Paleontology* **6**, 1-35.

Kale, H.W., II, and Maehr, D.S. (1990). "Florida's Birds." Pineapple Press, Sarasota, Florida.

Klein, R.G. (1980). The interpretation of mammalian faunas from stone-age archeological sites, with special reference to sites in the southern Cape Province, South Africa. *In* "Fossils in the Making" (A.K. Behrensmeyer and A.P. Hill, Eds.), pp. 223-246. University of Chicago Press, Chicago.

Kurtén, B. (1966). Pleistocene bears of North America. 1. Genus *Tremarctos*, spectacled bears. *Acta Zoologica Fennica* **115**.

Kurtén, B. (1984). Geographic differentiation in the Rancholabrean Dire Wolf (*Canis dirus* Leidy) in North America. *Carnegie Museum of Natural History Special Publication* **8**, 218-227.

Kurtén, B., and Anderson, E. (1980). "Pleistocene Mammals of North America." Columbia University Press, New York.

Lam, Y.M. (1992). Variability in the behaviour of Spotted Hyaenas as taphonomic agents. *Journal of Archaeological Science* **19**, 389-406.

Lee, D.S. (1969). Notes on the feeding behavior of cave-dwelling bullfrogs. *Herpetologica* **25**, 211-212.

Linhart, S.B., and Knowlton, F.F. (1967). Determining age of coyotes by tooth cementum layers. *Journal of Wildlife Management* **31**, 362-365.

Macdonald, D., ed. (1984). "The Encyclopedia of Mammals." Facts on File Publications, New York.

Marcus, L. (1960). A census of the abundant large Pleistocene mammals from Rancho La Brea. *Los Angeles County Museum, Contributions in Science* **38**, 1-11.

Martin, L.D., and Gilbert, B.M. (1978). Excavations at Natural Trap Cave. *Transactions of the Nebraska Academy of Sciences* **6**, 107-116.

Mead, J.I., and Van Devender, T.R. (1981). Late Holocene diet of *Bassariscus astutus* in the Grand Canyon, Arizona. *Journal of Mammalogy* **62**, 439-442.

Mech, L.D. (1970). "The Wolf: the Ecology and Behavior of an Endangered Species." Natural History Press, Garden City, New York.

Mellett, J.S. (1974). Scatological origin of microvertebrate fossil accumulations. *Science* **185**, 349-350.

Moon, E.L. (1940). Notes on hawk and owl pellet formation and identification. *Transactions Kansas Academy Sciences* **5**, 457-466.

Morgan, G.S. (1991). Neotropical Chiroptera from the Pliocene and Pleistocene of Florida. *Bulletin American Museum Natural History* **206**, 176-213.

Morgan, G.S., and Hulbert, R.C., Jr. (in press). Overview of the geology and vertebrate biochronology of the Leisey Shell Pit Local Fauna, Hillsborough County, Florida. *Bulletin of the Florida Museum of Natural History*.

Nowak, R.M. (1979). North American Quaternary *Canis*. *Monograph of the Museum of Natural History, University of Kansas* **6**.

Nowak, R.M., and Paradiso, J.L. (1983). "Walker's Mammals of the World, 4th ed. Vol. II." Johns Hopkins University Press, Baltimore.

Osmond, J.K., Carpenter, J.R., and Windom, H.L. (1965). Th^{230}/U^{234} age of Pleistocene corals and oolites of Florida. *Journal of Geophysical Research* **70**, 1843-1847.

Shipman, P. (1981). "Life History of a Fossil." Harvard University Press, Cambridge.

Skinner, J.D., and van Aarde, R.J. (1991). Bone collecting by Brown Hyaenas *Hyaena brunnea* in the central Namib Desert, Namibia. *Journal of Archaeological Science* **18**, 513-523.

Snyder, J.R., Herndon, A., and Robertson, W.B. (1990). South Florida rockland. *In* "Ecosystems of Florida" (R.L. Myers and J.J. Ewel, Eds.), pp. 230-277. University of Central Florida Press, Orlando.

Stafford, T.W., Jr., Hare, P.E., Currie, L., Jull, A.J.T., and Donahue, D.J. (1991). Accelerator radiocarbon dating at the molecular level. *Journal of Archaeological Science* **18**, 35-72.

Stock, C. (1942). Rancho La Brea: A Record of Pleistocene Life in California. *Los Angeles County Museum Science Series* **4**, 1-73.

Watts, W.A., and Hansen, B.C.S. (1988). Environments of Florida in the late Wisconsin and Holocene. *In* "Wet Site Archaeology" (B.A. Purdy, Ed.), pp. 307-323. Telford Press, Caldwell, New Jersey.

Webb, S.D., ed. (1974). "Pleistocene Mammals of Florida." University Presses of Florida, Gainesville.

Webb, S.D., and Wilkins, K.T. (1984). Historical biogeography of Florida Pleistocene mammals. *Carnegie Museum Natural History Special Publication* **8**, 370-383.

White, J.A., McDonald, H.G., Anderson, E., and Soiset, J.M. (1984). Lava blisters as carnivore traps. *Carnegie Museum Natural History Special Publication* **8**, 241-256.

Chapter 7

Human Impacts on the Environment during the Rise and Collapse of Civilization in the Eastern Mediterranean

STEVEN E. FALCONER* AND PATRICIA L. FALL†

*Department of Anthropology and †Department of Geography, Arizona State University, Tempe, Arizona 85287

INTRODUCTION

The legacy of human environmental impact in the Eastern Mediterranean dates back at least to the beginning of the Holocene. It has continued to the present in a variety of forms with a variety of causes. Paleoenvironmental, archaeological, and historical data reveal that this environmental deterioration has been discontinuous and multidimensional. The cultural roots for environmental degradation lie in ever-shifting patterns of human subsistence and settlement. The earliest impacts were intensive and localized, but they inspired cultural reponses that have, in turn, created today's extensively denuded landscapes.

Scientists and the lay public alike implicitly view the course of Near Eastern prehistory and history as a process of gradually developing agriculture, cities, and civilization amid landscapes with steadily increased erosion and decreased plant cover. Certainly the impoverished modern ecosystems of the Eastern Mediterranean provide expectable end products for such a view. However, this region achieved its spectacularly unnatural environment as the result of a series of human strategies with clear, sometimes spectacular, starts and stops.

This intermittent trajectory of human impact begins with the aggregation of Neolithic populations in very early, and ultimately abandoned, experiments in townlife. The effects of these nucleated communities on their surrounding landscapes necessitated new economic strategies, most notably agriculture and pastoralism, that have provided a persistent foundation for environmental impact to the present day.

The inception of complex urbanized society incorporating regional exchange of agarian commodities compounded these impacts through intensified orchard cultivation and forest clearance. Whereas subsistence farming and herding have been ever-present aspects of Eastern Mediterranean society since the Neolithic, the influences of urbanized civilization have been much less constant. The oscillating legacy of city life is reflected in paleoenvironmental signatures that are more varied and highly discontinuous.

Late Quaternary Environments and Deep History: A Tribute to Paul S. Martin, edited by David W. Steadman and Jim I. Mead. The Mammoth Site of Hot Springs, South Dakota, Inc. Scientific Papers, Volume 3. Hot Springs, South Dakota, 1995.

Thus, the course of social development in the Near East followed a frequently-changing path amid a constantly altered environmental setting. The modern biotic communities of this region are products of compound cultural influences, and may be profitably considered as much "humanized" as they are natural.

THE BEGINNINGS OF SEDENTISM, AGRICULTURE, AND PASTORALISM

The environment in which intensified human impact took place resulted from the climatic and vegetation changes at the end of the Pleistocene, approximately 10,000 yr B.P. Vegetation in Southwestern Asia changed from terminal Pleistocene steppe and desert-steppe conditions to forests and woodlands dominated by oak in the early Holocene. Pollen evidence from lake sediments in western Iran reveal that Pleistocene steppe gave way to an oak-pistacio forest-steppe beginning about 10,500 yr B.P.; tree cover became more dense in the early Holocene with the eventual establishment of the Zagros oak forest by 6000 yr B.P. (van Zeist and Bottema, 1982). This pattern of increasing forest expansion through the early Holocene is repeated in sediment cores from Lake Mirabad, also in western Iran (van Zeist, 1967), Lake Van in eastern Turkey (van Zeist and Woldring, 1978), and Sögüt Gölü in southwestern Turkey (van Zeist *et al.*, 1975).

Unlike the hill country of Iran and Turkey, where the Pleistocene vegetation is best described as steppe or desert-steppe, palynological data from the well-dated Lake Huleh core from northern Israel (Fig. 1) demonstrate that

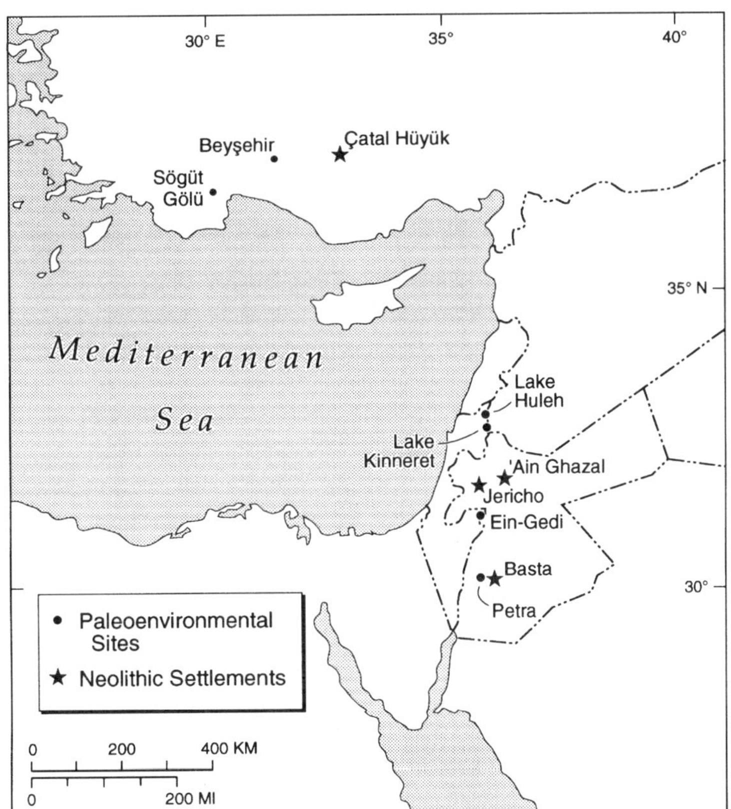

FIG. 1. The Eastern Mediterranean, showing the locations of selected Neolithic settlements (Çatal Hüyük, 'Ain Ghazal, Jericho and Basta) and paleoenvironmental sites (Sögüt Gölü, Beysehir, Lake Huleh, Lake Kinneret, Ein-Gedi and Petra).

a forest-steppe grew in the surrounding hills from 24,000 to 14,000 yr B.P. Oak forest subsequently expanded and became more dense between 14,000 and 10,000 yr B.P. (Tsukada *in* van Zeist and Bottema, 1982). This forest became more open with increased olive and pistachio at the expense of oak from 10,000 to 7400 yr B.P. (van Zeist and Bottema, 1982). This thinning and shift in forest taxa may implicate Neolithic clearance of oak stands.

A sediment core from the low elevation site of Ein-Gedi (approximately -300 m elevation) on the western shore of the Dead Sea documents an early Holocene steppe-forest with deciduous oak and pistachio trees along with steppe elements (*e.g.*, *Artemisia*) (Baruch, 1990)[1]. The vegetation around Ein-Gedi was most likely a desert-steppe that received arboreal pollen blown down from the Judean Mountains to the west. Thus, during the early Holocene the Levant supported expansive deciduous oak forests in the highlands, with more open oak forest-steppe and steppe vegetation in the lowlands.

A commonly-cited culprit for the earliest significant human impact on the early Holocene environment is the development of agriculture. However, some of the most persuasive evidence for early impact is more intimately associated with initial attempts at aggregated townlife and the associated clearance of natural vegetation for fuelwood and construction timbers. The prehistory of the Eastern Mediterranean features a number of sedentary communities that grew to unparalleled size during the eighth through sixth millennia B.C. (Rollefson and Köhler-Rollefson, 1989). Excavated plant macrofossils and animal bones suggest that wild resources played an important role in Neolithic subsistence (Rollefson and Simmons, 1986; Köhler-Rollefson, Gillespie, and Metzger, 1988). Some archaeologists infer management of wild game and long term experimentation with wild cereals (*e.g.*, Moore 1982), but in most cases these interpretations arise from limited botanical evidence and presuppose a lengthy period before animal husbandry resulted in morphological domestication.

A commonly asked question remains — why did humans turn so suddenly and irrevocably to their reliance on domesticated foods? Many scholars point to the dramatic climatic and vegetation changes at the end of the Pleistocene as the answer (*e.g.*, Butzer, 1975). It has been suggested that the forests and steppes were much richer in species during the Pleistocene, and thus, the wild precursors of domesticated cereals were limited by competition within this more diverse vegetation (Moore, 1982). It was left for humans to clear the Mediterranean forests to make way for their beloved cereal crops. It now appears more likely that the aggregation of relatively large Neolithic settlements (*i.e.*, with several thousand occupants) initiated intensive exploitation of local environments. In the aftermath of plant and animal domestication, natural resources (game, wild plant foods, and wood) became depleted around these "proto-towns," forcing their inhabitants to dedicate themselves increasingly to intensified agriculture in seriously degraded local environments.

Environmental impact, and the rise and collapse of Neolithic towns

Several Neolithic settlements are particularly noteworthy because of their anomalously large sizes and relatively rapid abandonments. They incorporate many trappings of early urbanism, yet Eastern Mediterranean urban traditions did not become permanently established until two to three millennia after their collapse. These enigmatic settlements include Anatolian examples like Çatal Hüyük, which covered 13 hectares and is renowned for its formal religious architecture (Mellaart, 1967), and several Levantine sites such as Jericho, where excavations revealed a Pre-Pottery Neolithic town wall and defensive tower (Kenyon, 1960; Bar-Yosef, 1986) and Basta, a settlement of at least 10 hectares with well-preserved multi-storey architecture (Gebel *et al.*, 1988; our Fig. 1 for site locations). The fate of these proto-towns remains unexplained in general terms. Among these sites only 'Ain Ghazal provides sufficiently detailed environmental data from which we may formulate a general explanatory model for the inception of human alteration of the Near Eastern landscape. 'Ain Ghazal demonstrates that severe environmental impacts followed closely on the heels of early agriculture and may have forced the abandonment of Neolithic town life (Rollefson and Köhler-Rollefson, 1989).

'Ain Ghazal was occupied in the Pre-Pottery Neolithic and Pottery Neolithic periods (*ca.* 9200-7700 yr B.P.) during which its population increased six fold, reaching 12.5 hectares and an estimated 3600 inhabitants in Pre-Pottery Neolithic C (Köhler-Rollefson and Rollefson, 1990). Faunal and architectural data provide several lines of evidence for systematic alteration of the plant and animal communities around 'Ain Ghazal. Bone assemblages demonstrate that a wide variety of wild game was available from diverse habitats (*e.g.*, forest, woodland, steppe, and wetland). The frequencies of wild forest species (*e.g.*, deer) dwindle through the Pre-Pottery Neolithic B and C, and Pottery Neolithic in favor of increasingly abundant domesticated species, especially goat (Köhler-Rollefson and Rollefson, 1990).

Architectural data specify some causes for the systematic deforestation that helped precipitate this shift in animal exploitation. Early Pre-Pottery Neolithic houses incorporated large oak construction timbers to support roofs and walls. During the Pottery Neolithic, 'Ain Ghazal's houses utilized progressively smaller timbers, and finally omitted their use altogether by the end of the site's occupation (Rollefson *et al.*, 1992; Banning and Byrd, 1984; 1987). A similar pattern of timber use, with much greater potential for deforestation, is implicated by the heavy use of lime plaster. Each Pre-Pottery Neolithic house was founded on plaster floors and lower walls that were refinished repeatedly. This plaster was produced by slaking limestone, which is abundantly available from local bedrock outcrops, at high temperature to produce quicklime. A coat of plaster for an average house required approximately 3.3 tons of quicklime, and the production of each ton of quicklime consumed approximately four tons of fuelwood. Thus, a single episode of plastering for one house removed the equivalent of six mature oak trees from 'Ain Ghazal's surroundings (Rollefson and Köhler-Rollefson, 1992). 'Ain Ghazal may have included 600 households by 7700 yr B.P. Given that most houses were replastered repeatedly, the potential for local deforestation is staggering (Rollefson and Köhler-Rollefson, 1992; our Table 1).

Just as construction timbers dwindled and disappeared during 'Ain Ghazal's occupation, so did domestic plastering. Over the course of

TABLE 1. Cumulative architectural tree consumption at 'Ain Ghazal, Jordan (After Köhler-Rollefson and Rollefson, 1990).

Years B.P.	Trees Consumed	Depleted Land (ha)
7700	57,200	3,268
8200	35,850	2,049
8700	12,230	669
9200	1,260	72

the Pre-Pottery Neolithic plaster floors became progressively thinner, and by the Pottery Neolithic floors were left unplastered altogether (Banning and Byrd, 1984). The similar, and undoubtedly related, patterns of diminished use of construction timbers and domestic plastering strongly imply a correspondingly diminished availability of oak trees. Thus, a variety of architectural changes at ‘Ain Ghazal reflect the impact of deforestation on the structural integrity of Neolithic communities.

The beginnings of pastoralism

While the environmental impacts around ‘Ain Ghazal and other Neolithic towns must have been drastic, they would have remained localized. However, the cultural response to costs and loss of travel time to distances of three to four kilometers (Chisholm, 1968). These same data suggest that such travel requirements eventually would lead to the establishment of seasonal or permanent outlying herding camps. Indeed, by the end of its occupation, ‘Ain Ghazal’s faunal assemblage suggests local consumption from a residual herd, while the vast majority of goats were grazed at considerable distances (Köhler-Rollefson and Rollefson, 1990; Pires-Ferreira, 1975-77). Thus, intensive local agriculture and its attendant deforestation inspired the seasonal and full-time pastoralism so characteristic of ancient and modern Eastern Mediterranean society. Rather than a precursor of settled agriculture, mobile pastoralism is best interpreted as a reaction to it (Hole, 1974). This response entailed new, widespread

TABLE 2. Cumulative impact of agriculture, grazing, and wood harvesting at 'Ain Ghazal, Jordan (After Köhler-Rollefson and Rollefson, 1990).

Years B.P.	Exploited Land (ha)	Average Distance (km)
7700	17,100	7.4
8200	14,000	6.6
8700	4,600	3.8
9200	2,020	2.5

this altered landscape insured that human impact would become a permanent factor across the entire Middle East. That response involved the interrelated development of sedentary agriculture and non-sedentary pastoralism.

As wild resources became increasingly distant, the inhabitants of ‘Ain Ghazal dedicated the deforested lands immediately surrounding them to pasturage and plant cultivation. It is estimated that this continued intensive local exploitation soon required that goats be taken at least seven kilometers to browse (Köhler-Rollefson and Rollefson, 1990; our Table 2). Ethnographic data suggest that traditional agriculturalists can accomodate daily transport potential for human impact on the environment in form of defoliation from browsing, particularly by goats.

The stage was now set for extensive defoliation in the wake of pastoralists and their herds traversing vast expanses of forest-steppe, steppe, and desert terrain. Herders constitute a characteristically fluid aspect of traditional societies, both ancient and modern, in Southwestern Asia. They are particularly noteworthy for their resilience in the face of political and economic pressures (Adams, 1978). During periods of instability or excessive demands (*e.g.*, taxes, corvée labor, military conscription), families or whole communities may shift from settled farming to mobile pastoralism. Ironically, pastoralism and its associated impacts became more pronounced in the face of both maximum and minimum central authority, as seen during periods of pronounced state bureaucracies (*e.g.*, in the Ottoman Empire) and their collapse (*e.g.*, during the Middle Babylonian period in Mesopotamia [Adams, 1981] or Early Bronze IV in the Southern Levant [Dever, 1980]).

ARBORICULTURE AND ANTHROPOGENIC INDICATORS OF FOREST CLEARANCE

The detrimental effects of excessive pastoralism are not always easy to discern palynologically, even for historical periods in which they are well-documented otherwise. However, there are a number of "anthropogenic indicators" (Behre, 1981; 1990; Bottema, 1982[1985]; Bottema and Woldring, 1990) that include plant taxa which thrive in the wake of human disturbance of soils and competing flora. In addition to the pollen of cultivated taxa (*e.g.*, Cerealia), *Plantago lanceolata*-type and *Sarcopoterium*-type pollen are good secondary anthropogenic indicators. *Plantago lanceolata* is a characteristic component of vegetation in disturbed soils, including heavily grazed areas (Behre, 1990). *Sarcopoterium* is a successional plant that spreads after degradation or destruction of forests or shrublands. Animals dislike its taste, and it commonly thrives in abandoned fields or orchards in the Near East. Other indicators of disturbance vegetation include Gramineae, *Artemisia*, and weed taxa (*e.g.*, Liguliflorae, Tubuliflorae and Chenopods). Grass pollen generally declines with increased grazing. *Artemisia* is a common element of open steppe vegetation that may result from drier climatic conditions, as well as human disturbance. Similarly, chenopods increase in more open vegetation, naturally and through human disturbance. Both Tubuliflorae and Liguliflorae are favored by open habitats. Throughout the eastern Mediterranean, these anthropogenic indicators first become sharply pronounced in the mid-Holocene (Bottema and Woldring, 1990; Baruch, 1990; Behre, 1990).

During the middle and late Holocene the development of regional economic exchange networks altered the expression of human impact on the environment. This is particularly well demonstrated by pollen signatures for forest clearance. Changes in the composition of arboreal vegetation often implicate human management of specific tree taxa or wholesale forest clearance. At Beyehir Gölü, Turkey, pollen spectra reveal that an early-mid Holocene oak-pine-juniper forest gave way to a variety of anthropogenic vegetation types beginning *ca.* 3000 yr B.P. (Bottema and Woldring, 1990). This shift is ascribed to forest clearance for farming and orchard management over a period of 1000-2000 years (Roberts, 1990). Subsequently, reduced human habitation in the region permitted a regeneration of forest, now dominated overwhelmingly by pine. Particularly clear anthropogenic signatures also are apparent in the pollen diagrams from Sögüt Gölü and Koiladha (also in Turkey), which show dramatic increases in olive (*Olea*), walnut (*Juglans*) and manna-ash (*Fraxinus*), a shift from deciduous oak (*Quercus cerris*-type) to evergreen oak (*Quercus calliprinos*-type or *Q. cocciferra*-type), and a decrease in the ratio of arboreal to non-arboreal pollen types (Bottema and Woldring, 1990) (Fig. 2). Several authors collectively ascribe these intervals of human impact to a common "Beysehir Occupation Phase" in the Eastern Mediterranean between ca. 3500 and 1500 yr B.P., despite minimal chronological data for most of the cores in question (van Zeist *et al.*, 1975; Bottema and Woldring, 1990; Behre, 1990).

Pronounced spikes in *Olea* pollen frequencies dating to *ca.* 3000 yr B.P. in cores from

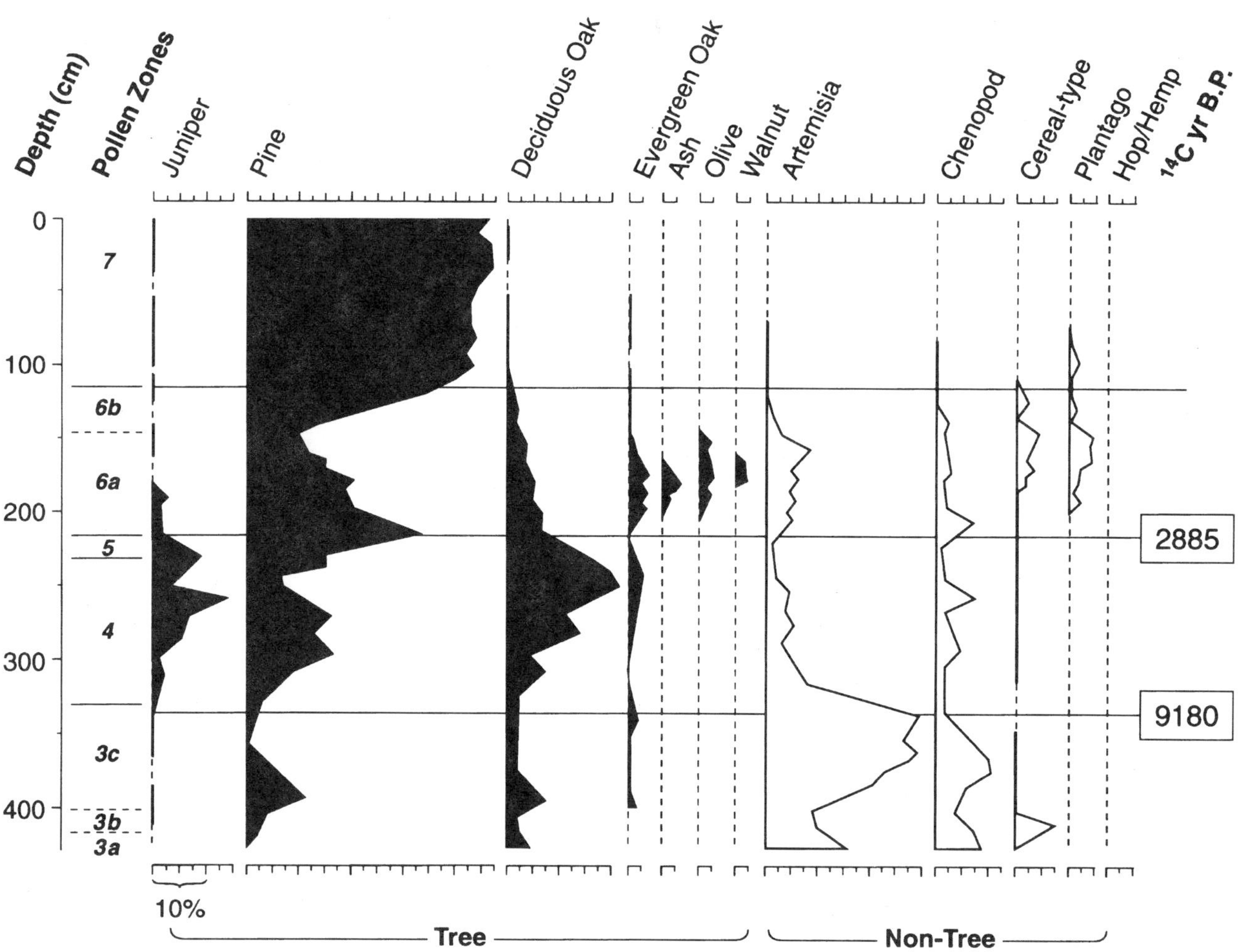

FIG. 2. Pollen diagram from Sögüt Gölü southwestern Turkey, reflecting vegetation change from late-glacial steppe in the terminal Pleistocene to mixed forests of the Holocene. Beysehir Occupation Phase showing human manipulation of forest trees and cultivation of cereals in the middle and late Holocene. After Roberts (1990); data from van Zeist *et al.* (1975).

Turkey and Greece (*e.g.*, Sögüt Gölü, Koiladha) (Fig. 2) probably reflect the impacts of population growth and cities, along with their extensive fruit orchards, during the Iron Age and classical periods. Greek and Anatolian urbanism and regional exchange created large-scale demands for commodities such as olive oil, which could be produced, packaged, and shipped in bulk. The pollen data show that these conditions led to forest clearance to permit the planting of substantial olive orchards. This intrusion into wild forests led to various alterations of wild forest associations. Several cores show a gradual shift from deciduous to evergreen forms of oak. This shift is accompanied by dramatic increases in pine and fir at Koiladha, increased juniper at Beysehir Gölü, and drastic decreases in pine and juniper at Sögüt Gölü (Bottema and Woldring, 1990; Roberts, 1990).

Whereas localized Neolithic environmental manipulation appears most clearly in changing floral and faunal assemblages from specific sites, the effects of early urbanized societies are much broader in scope and often emerge from comparative analyses of oscillations in regional plant communities. A relatively detailed portrait of the impacts of early civilization on the vegetation of the Southern Levant may be discerned from a variety of archaeological, historical, and palynological sources. This portrait reveals a suite of human influences, some relatively modest and gradual, others highly pronounced and punctuated. When viewed jointly, they provide a paleoenvironmental reflection of the economic fortunes and misfortunes of this corner of the Eastern Mediterranean.

REGIONAL ECONOMIES AND THEIR ENVIRONMENTAL IMPACTS IN THE SOUTHERN LEVANT

Paleobotanical and palynological data from the Southern Levant are particularly interesting because they record multiple episodes of human intervention that can be interpreted in light of well-established archaeological and historical data for regional populations and settlement patterns. We define the Southern Levant as the lands within modern Israel, the West Bank, and western Jordan. Following the early, ultimately abandoned, proto-towns of the Pre-Pottery Neolithic, Levantine populations inhabited more dispersed agricultural hamlets during the Pottery Neolithic and the subsequent Chalcolithic Period. These latter two periods are a formative prelude to the development of cities, formal political authority, and regional economic exchange during the Bronze and Iron ages of the fourth through first millennia B.C. (Table 3). The Bronze Age, which is associated with the Canaanites of the Old Testament and other historical sources, is marked by two peaks in regional population density and urban development: Early Bronze II-III (*ca.* 2700-2300 B.C.) and Middle Bronze IIB-C (*ca.* 1800-1500 B.C.) (Richard, 1987; Dever, 1987). The Late Bronze Age, which featured extensive long-distance exchange of manufactured goods in the Eastern Mediterranean (*e.g.*, pottery), was nevertheless a period of reduced sedentary population and urban decline in the Southern Levant (Gonen, 1984; Leonard, 1989). The Iron Age witnessed the development of the Israelite kingdoms, but never developed the population densities or large urban centers of the Bronze Age (Broshi, 1978; Fritz, 1987).

Following the foreign political domination of the Persian and Hellenistic periods, population growth and economic development were renewed on an unprecedented scale once the Southern Levant was incorporated into the Roman and Byzantine empires (Broshi, 1979). Subsequent Islamic political authorities

TABLE 3. Post-Neolithic archaeological chronology for the Southern Levant.

Period(s)	Duration
Islamic/Modern	A.D. 638-present
Roman/Byzantine	63 B.C.-A.D. 638
Persian/Hellenistic	587-63 B.C.
Iron Age	ca. 1200-587 B.C.
Late Bronze Age	ca. 1500-1200 B.C.
Middle Bronze Age	ca. 2000-1500 B.C.
Early Bronze Age	ca. 3500-2000 B.C.
Chalcolithic	ca. 4500-3500 B.C.

(through the Ottoman Empire) invested less effort toward the economic well-being of the Southern Levant, and the region's population declined drastically and commonly became mobile pastoralists. Only since the First World War has the regional population level surpassed that of the classical periods (Broshi, 1979).

The advent of Levantine orchard cultivation

Roughly three millennia after the collapse of Neolithic proto-towns, the emergence of towns and cities amid regional economic networks in the late fourth and early third millennia B.C. involved the clearance of oak forests for the purpose of planting orchard crops, including grape vines, date palms, fig trees and, especially, olive trees (Zohary and Spiegel-Roy, 1975; Stager, 1985; Neef, 1990). Although pollen cores do not show increased frequencies for these taxa until *ca.* 2000 B.C., the earliest macrofossils have been excavated from Levantine Chalcolithic sites dating substantially earlier.

Neef (1990) suggests that olive cultivation started earliest in Northern Palestine/Southern Syria. In many cases it is unclear whether the Chalcolithic samples are domesticated or wild because much of the evidence consists of *jift*, the crushed remains of olive pits left after pressing. Indeed, wild olive still persists in modern Jordan as a minor constituent of forests dominated by deciduous oak (*Q. ithaburensis*). It is unclear, however, if these modern trees are relics of a more expansive olive forest or merely escaped cultivars. The most detailed evidence for olive cultivation in the Chalcolithic comes from three sites in the Jordan Valley. At Abu Hamid and Tell esh-Shuna [North] the vast majority of excavated wood samples are olive, many of which come from ash deposits, showing a pronounced use of olive for fuelwood (Dollfus *et al.*, 1988; Neef, 1990: Table 2). At Teleilat Ghassul, as well as Abu Hamid, olive wood beams also were used for construction (Hennessy, 1969; Dollfus *et al.*, 1988). This evidence reflects intensive olive cultivation for fuel, construction, and oil production. (Pickling of olives was not introduced until the Hellenistic and Roman periods following the fourth century B.C. [Borowski, 1987]). These remains probably came from domesticated orchards, since it seems unlikely that olive timber would have been harvested selectively from mixed oak-olive forests. It may be inferred that by the Chalcolithic much oak forest had been cleared or at least severely reduced in extent. Therefore, calibrated radiocarbon dates from wood fragments excavated from these three sites in the Jordan Valley suggest the first cultivation of extensive olive orchards *ca.* 4500-4000 B.C. (Neef, 1990: Table 4).

Palynological spectra from Ein-Gedi

Two pollen cores are particularly illustrative of human environmental impact in the Southern Levant. The first, taken along the western shore of the Dead Sea near Ein-Gedi, spans at least the last 8000 years of the Holocene, based on two radiocarbon dates toward the core's base (Fig. 3). The uppermost one meter of the core was truncated during recovery and published data cover the lower five meters only. A supplementary core of the top meter has been taken from the same locality but has not been published (Baruch, 1990).

The Ein-Gedi core has been divided into four pollen assemblage zones (Baruch, 1990; Fig. 3). The earliest zone, Zone X, documents forest vegetation dominated by deciduous oak (*Q. boissieri*-type), with a modest element of pistachio. *Artemisia* occurs at substantial frequencies through Zone X and into the subsequent Zone Y. This anthropogenic indicator seems not to reflect forest clearance for orchard cultivation, since there is little change in forest taxa, and *Olea* remains at very low levels through Zone X. Instead, it may represent an

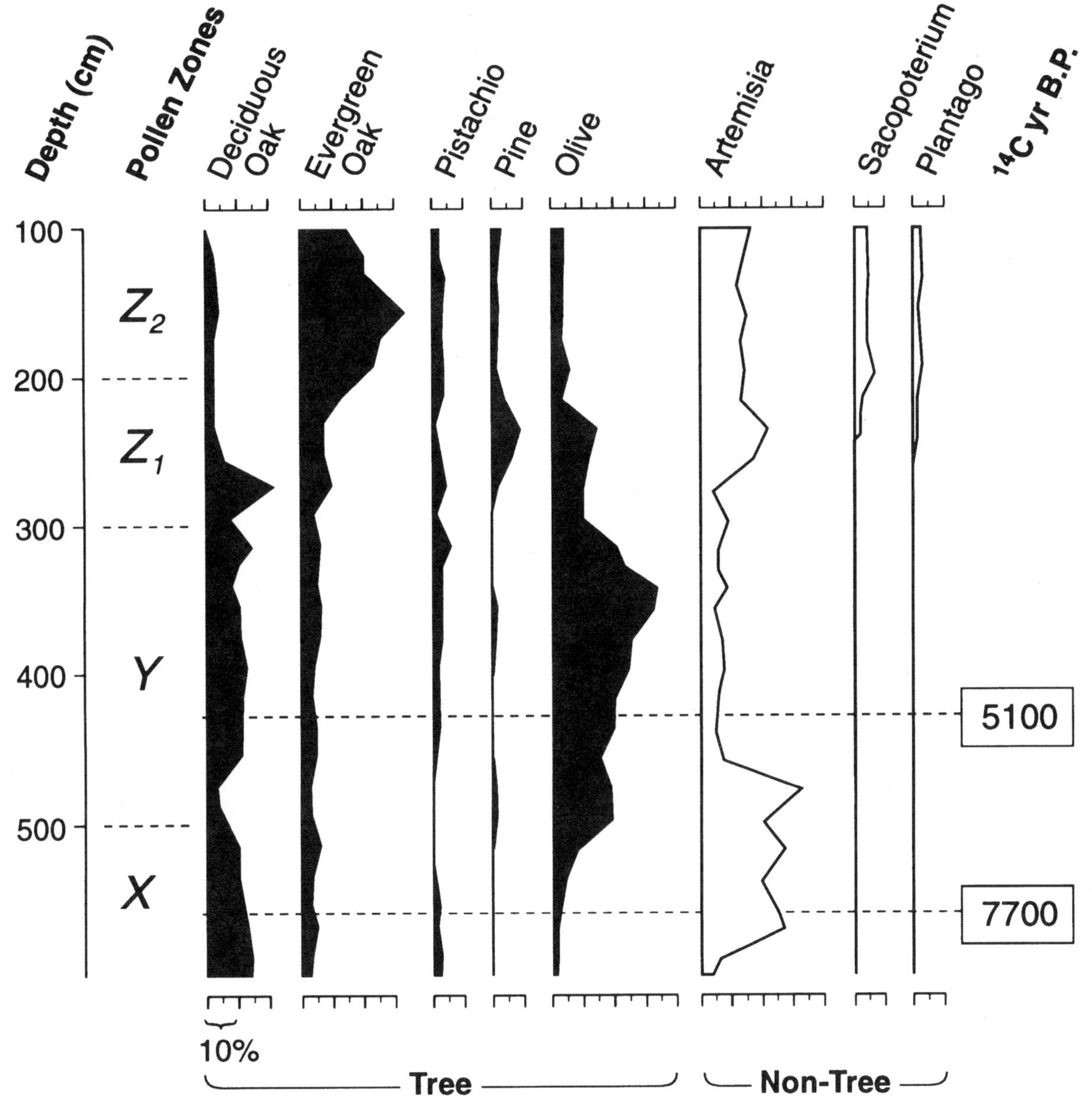

FIG. 3. Pollen diagram from Ein-Gedi, Dead Sea, Israel. After Baruch (1990). Based on correlation with Lake Kinneret, Baruch (1990) places the basal age of this core at approximately 5600 yr B.P. with the maximum *Olea* pollen peak corresponding to the Roman/Byzantine period, rather than the Bronze Age as interpreted in this paper based on the existing radiocarbon chronology.

early Holocene remnant of late Pleistocene steppe vegetation or, more interestingly, it may reflect another form of impact, such as substantial grazing. Based on the core's lower radiocarbon date, Zone X corresponds roughly with the Late Neolithic, and it may provide an early palynological indication of human impact following the inception of pastoralism.

Zone Y continues to indicate an oak-pistachio forest, now accompanied by elevated *Olea* values during the Chalcolithic and Bronze ages. Baruch (1990) suggests that olive cultivation was concentrated in lower forest areas, since frequencies of upper elevation deciduous oak remain high. Zone Y provides palynological corroboration for the contention of Zohary and Spiegel-Roy (1975) and Neef (1990) that olive cultivation began in the Chalcolithic Period. However, in the region west of the Dead Sea this cultivation apparently involved minimal forest clearance. Arboreal pollen frequencies increase steadily until the upper reaches of Zone Y, primarily due to increased *Olea*, but other arboreal taxa certainly hold their own.

After dropping at the top of Zone Y, olive frequencies remain at about 10-15 % in Zone Z_1. Based on the average sedimentation rate for the upper core, this zone corresponds roughly to the Roman and Byzantine periods. These *Olea* values do not approach those of the preceding Chalcolithic/Bronze Age peak, but they are accompanied by several significant new trends. First, a major shift in forest composition is signalled in the middle of Zone Z_1. Deciduous oak frequencies drop, leading to an immediate jump in pine values. Baruch (1990) notes that pine (*Pinus halepensis*) thrives when competition from other arboreal taxa is lacking. Following a relatively brief pine spike, forest vegetation becomes dominated by evergreen oak (*Q. calliprinos*-type). This major shift in forest composition is intriguing because it lags behind the earlier peak in *Olea* frequencies. Although olive production may have been less intensive than in earlier periods, orchards now apparently moved up in elevation and displaced natural stands of deciduous oak. Alternatively, this shift may represent a cumulative effect of orchard development over several millennia, which finally triggered a wholesale alteration of wild forest composition. A second expression of human impact is signalled in the later samples of Zone Z_1 by elevated *Artemisia* values and the first appearance of *Sarcopoterium*-type and *Plantago lanceolata*-type in significant frequencies.

During Zone Z_2, covering the first centuries of Muslim rule in the Southern Levant, evergreen oak forest is the dominant arboreal vegetation. Interestingly, this evergreen oak replaces the original deciduous oak (*Quercus boissieri*) in the forests of the Levant following the abandonment of wide-spread olive cultivation. The multiple stemmed evergreen oak (*Quercus calliprinos)* is better adapted to human use and accepts browzing and wood cutting better that the single trunk deciduous oak (*Quercus boissieri*) (Shmida, 1980; Baruch, 1990). Although olive cultivation may have been largely abandoned, continued human impact appears to be signalled by the persistence of two trends from upper Zone Z_1: elevated *Artemisia* values and considerably enhanced frequencies of *Sarcopoterium*-type and *Plantago lanceolata*-type pollen. These trends probably reflect greater proportions of Levantine populations opting for pastoralism.

Palynological spectra from Lake Kinneret

The second core, taken from the southeastern portion of Lake Kinneret (Sea of Galilee), is subdivided into six pollen assemblage zones that span the second half of the Holocene (Baruch, 1986; Baruch, 1990: fig. 2) (Fig. 4). The two lowest zones show an aggregate trend of decreased arboreal pollen, suggesting forest clearance. In Zone X_1, at the core's base, dense oak forest recedes briefly due a short-term increase in *Olea* roughly contemporaneous with the population growth and early urbanism of Early Bronze II-III. Following a regeneration of oak forest during the collapse of Early Bronze urbanism, dwindling oak frequencies and a another modest olive peak (*ca.* 3000 B.P.) in Zone X_2 implicate renewed forest clearance and orchard development during the Middle Bronze, Late Bronze, and Iron ages. The anthropogenic indicators *Sarcopoterium*-type and *Plantago lanceolata*-type, previously quite rare, also show very modest increases through the zone.

Zone Y then presents perhaps the most graphic palynological illustration of human intervention anywhere in the Eastern Mediterranean. While olive frequencies rapidly approach 60 % of the pollen sum, oak frequencies recede to roughly 5 %. Among other anthropogenic indicators, an abrupt rise in *Vitis* (grape) pollen coincides with that of *Olea*, and *Sarcopoterium*-type and *Plantago lanceolata*-type jump to much higher values (*ca.* 5 %) during and after this peak. The apparently modest grape pollen peaks are noteworthly, since *Vitis* produces very low amounts of pollen. A radiocarbon date of 2145±175 yr B. P. immediately above the olive peak shows that these patterns reflect an era of large-scale olive oil production in Northern Palestine during the Roman and Byzantine periods. Indeed, Latin records mention olives, grapes, and figs as the three major crops of the region (Felix, 1982 *in* Baruch, 1990). Levantine olive oil and wine, well-regarded in the Eastern Roman Empire, found receptive markets, particularly in Egypt.

Almost as abruptly as they rose, *Olea* frequencies dwindle into zones Z_1, Z_2 and Z_3, which date roughly from the Muslim conquest of Palestine to the present day. These zones show a regenerated forest quite different from the more pristine forest of Zone X_1. While deciduous oak values never attain pre-classical levels, evergreen oak and pistacio become more pronounced elements than ever before. *Sarcopoterium*-type and *Plantago lanceolata*-type, once quite rare, now are common in the regional pollen spectra.

The Kinneret and Ein-Gedi cores reflect a legacy of regional human impact from the mid-Holocene to the present day. Before and during the classical periods human environmental manipulation primarily involved orchard cultivation at the expense of natural forests. Despite the decimation of wild tree cover, these forests did regenerate, although in significantly different forms. Interestingly, the rise in anthropogenic shrubs and herbs, which might be attributed to overgrazing by pastoralists and their flocks, begins at the height of Roman arboriculture. During subsequent periods, in which we know pastoralism increased, pollen frequencies for these disturbance taxa (e.g., *Sarcopoterium*-type and *Plantago lanceolata*-type) remain consistent, while those for wild forest taxa rebound. Thus, regional deforestation had its roots in the production of trade commodities by urbanized populations. Further, those impacts that might be attributed to pastoralism arose during the apex of Roman imperial authority. Therefore, pastoral impacts may be responses to central authority, as well as products of its collapse.

Hyrax middens from southern Jordan

Another body of paleobotanical data, in the form of pollen and plant macrofossils from hyrax middens, corroborates many of the trends seen in the Ein-Gedi and Kinneret pollen cores. In a manner similar to that of packrats (*Neotoma* sp.) in the American southwestern deserts, the Syrian rock hyrax (*Procavia capensis*) constructs middens composed of numerous plant fragments, fecal pellets, and faunal parts. These deposits include remnants of browsing by the hyrax, which "will feed on almost any plant" (Harrison, 1968:320), preferably those near where it lives. Hyrax

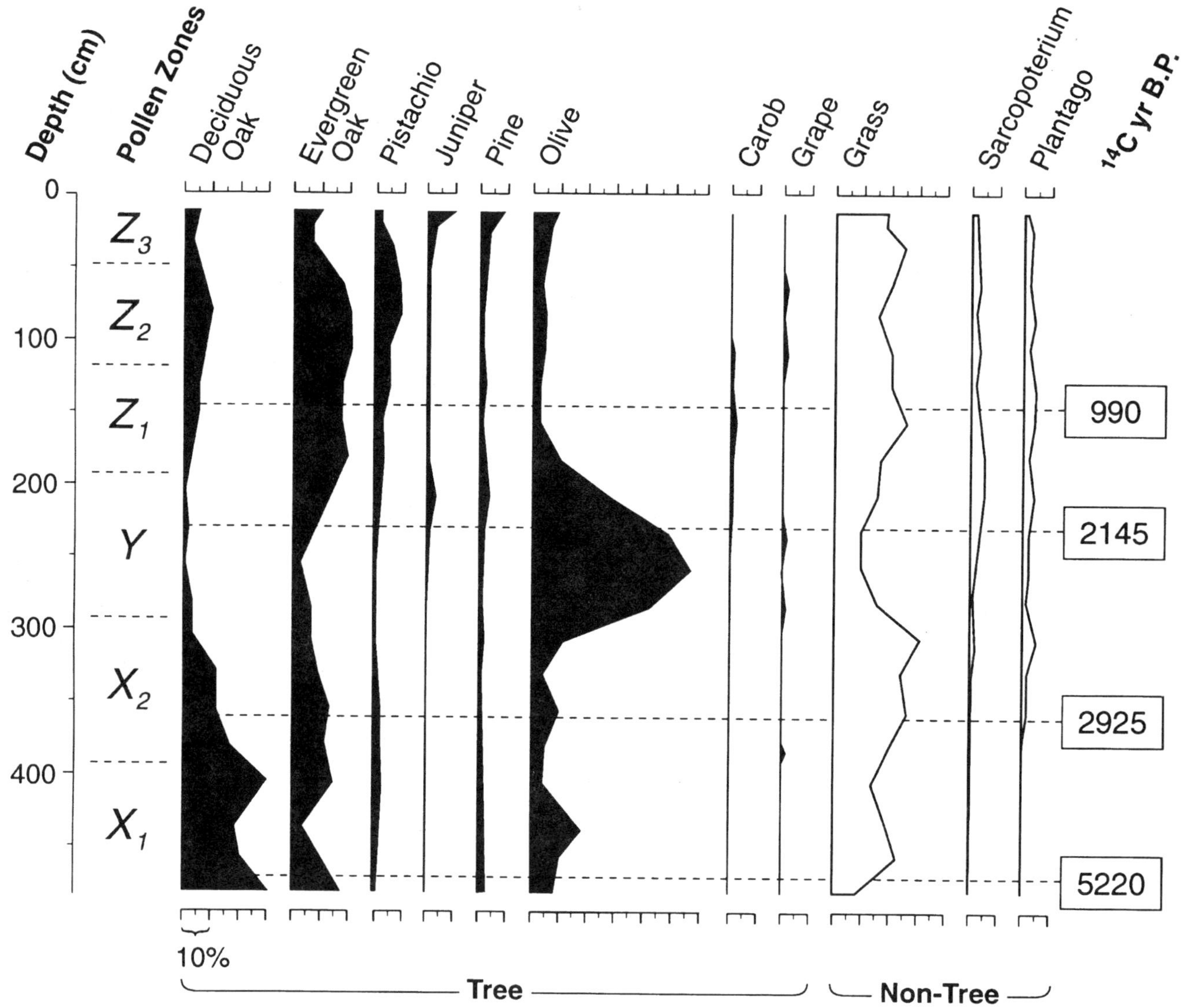

FIG. 4. Pollen diagram from Lake Kinneret, Sea of Galilee, Israel. After Baruch (1990).

middens, which can be quite large (*i.e.*, more than 50,000 cc), are hardened and preserved by hyracium, thc crystallized urine of the hyrax.

A series of hyrax middens has been collected from mountain crevices near the ancient city of Petra in southern Jordan (Fall, 1990; Fall *et al.*, 1990). Three hardened, or indurated middens have been radiocarbon dated. An unindurated midden and pollen from a surface soil sample provide modern analogs. The indurated middens relate to periods of distinctly different local economy and land use (Fall *et al.*, 1990). When calibrated to calendric years (following Pearson, 1983), radiocarbon dates place two middens (Petra 6 and Petra 9) in the Roman and Byzantine periods. However, since their 95 % confidence intervals overlap, these middens are used jointly to interpret classical period vegetation. The calibrated date for Petra 2 places it in the Early Islamic Period, while the unindurated midden (Petra 1) is considered modern, since it contains pollen from *Eucalyptus*, a tree introduced in the twentieth century.

Comparison of total arboreal pollen frequencies in middens 6 and 9 with surface pollen spectra from Syria and Lebanon (Bottema and Barkoudah, 1979) suggests that the woodlands around Petra already had become a degraded Mediterranean steppe-forest by the Roman and Byzantine periods (Fig. 5). Sparse oak and pistachio were accompanied by juniper on lower mountain slopes, and intensively managed olive orchards. Shrub undergrowth consisted of *Artemisia*, grass, *Ephedra*, and *Sarcopoterium*. Subsequently, total arboreal pollen drops drastically in the Early Islamic and modern middens to levels analogous with those of modern desert-steppe vegetation. All of the major arboreal taxa begin with modest frequencies and show a general decline over the interval represented by these middens. Trees surviving in restricted habitats were accompanied by sparse woody shrubs, while steppe and desert species came to dominate the "natural" vegetation. This pervasive decrease may be the legacy of several centuries of wood harvesting to supply a substantial city population with building material and fuel. Although it does not decline as rapidly as other arboreal taxa, the drop in *Olea* implicates a gradual abandonment of orchards by the Early Islamic Period. Although *Sarcopoterium*-type is present in low percentages (< 1 %) in these spectra, *Plantago lanceolata*-type follows the same general pattern seen in the Kinneret and Ein-Gedi cores by peaking during the latter part of classical periods, suggesting the possibility of increased pastoralism during imperial rule. Several other desert taxa or disturbance indicators (*e.g.*, goosefoot-amaranth, nettle, and umbel) peak during the Early Islamic Period, as reflected by midden 2 (Fig. 6). The modern midden (midden 1) shows a further increase in caperbush, a toxic shrub, and in weedy disturbance taxa, including borage and Liguliflorae. These patterns jointly suggest a major increase in pastoralism following the collapse of Byzantine rule, which reduced already degraded forests to the mere remnants of oak-pistachio woodland and denuded steppe that characterize the region today.

Popular conceptions of environmental degradation in the Eastern Mediterranean tend to emphasize the effects of sheep and goat pastoralism following the advent of Islam. Paleobotanical data from the Southern Levant reveal a much longer legacy in which pastoralism was simply one of many compounded influences. In particular, these data show that the development of urbanism and regional economies over the last five millennia engendered most of the widespread deforestation that epitomizes human impact on the Eastern Mediterranean landscape. The archaeological and paleobotanical records illustrate that processes of both urbanization and deforestation are strikingly reversible. Episodes of nucleated settlement and centralized authority in the Bronze and Iron ages, and during the classical periods were set off by equally significant intervals of urban economic and political collapse. Forest vegetation impoverished by the demands of urban populations *did* regenerate when these demands were eased, but never in its original form. Thus, even those pockets of woodland visible today are less remnants of pristine natural vegetation than they are artifacts of human intervention.

CONCLUSIONS

The earliest evidence for human environmental impact in the Eastern Mediterranean comes from some of the world's first experiments in aggregated settlement. In these cases, localized depletion of wild resources apparently necessitated the development of sedentary farming, which in turn inspired a comple-

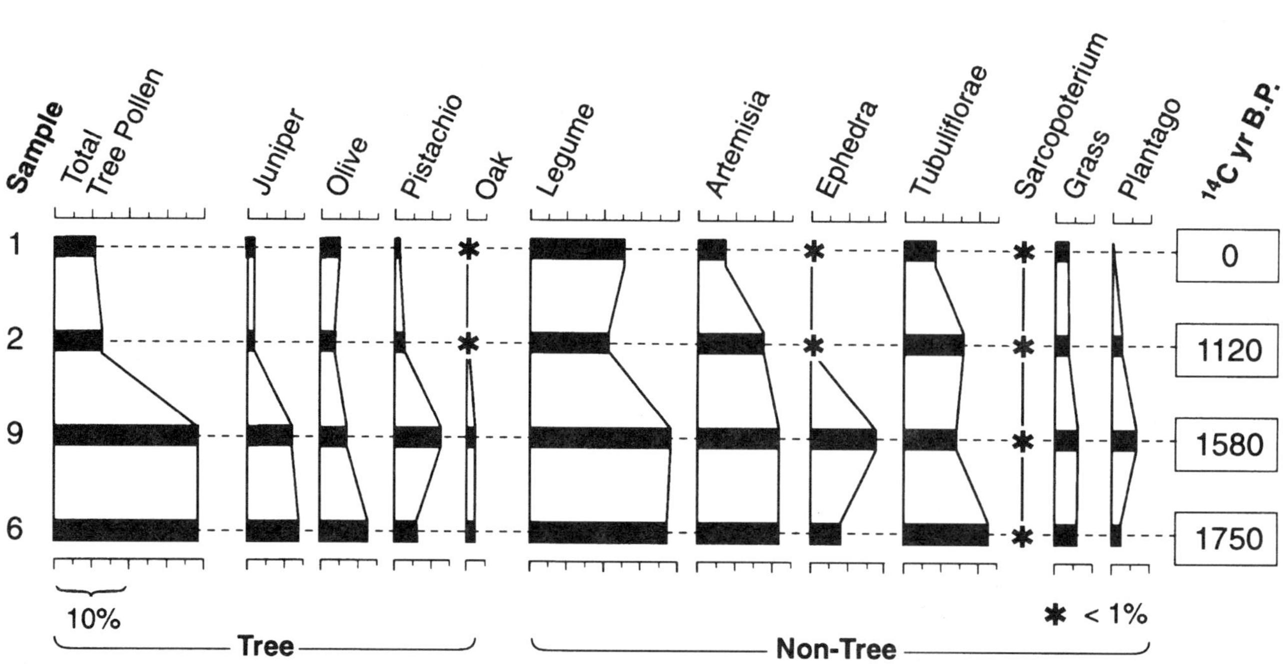

FIG. 5. Pollen from hyrax middens, Petra, Jordan. Tree and non-tree pollen taxa showing significant decreases from the Roman/Byzantine period (middens 6 and 9) to the Islamic and modern periods (middens 2 and 1). After Fall (1990).

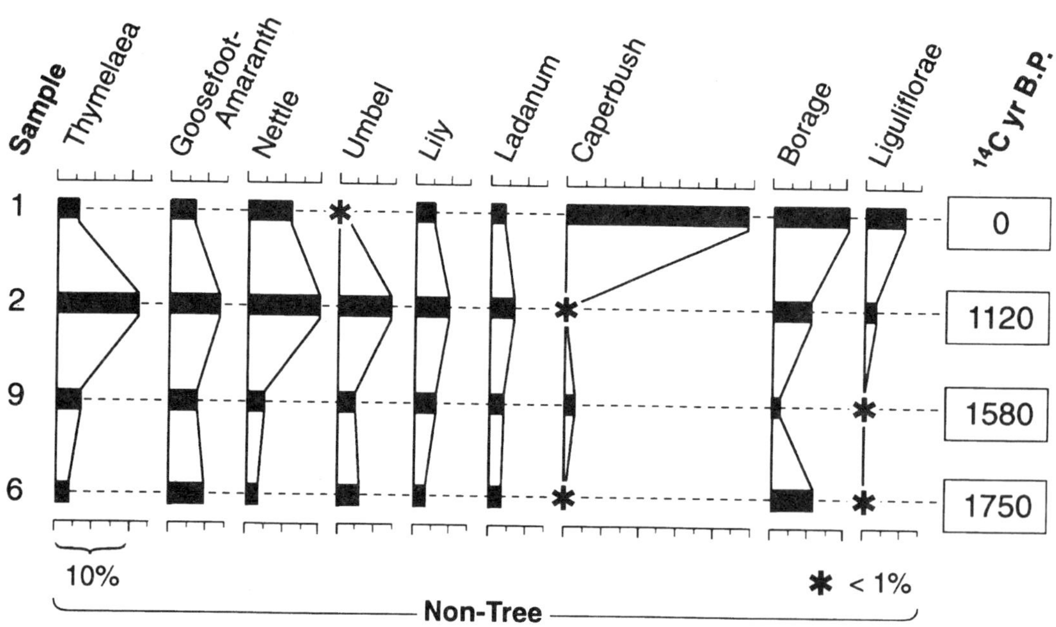

FIG 6. Pollen from hyrax middens, Petra, Jordan. Shrub and herb pollen taxa showing significant increases following the Roman/Byzantine period (middens 6 and 9). After Fall (1990).

mentary strategy of non-sedentary animal husbandry. Archaeological and paleobotanical data suggest that once these domestication economies were embraced, significant human impact on the environment was assured and there was no going back to pre-agricultural lifestyles.

Environmental impact expanded to a regional scale in the middle and late Holocene with extensive forest clearance associated with the rise of civilization and regional exchange of agricultural commodities. In addition to grain agriculture, orchard cultivation of grape, fig, and especially olive figured prominently as economic mainstays of the Eastern Mediterranean, first in the Bronze and Iron ages and later, on a massive scale, during the heyday of Roman/Byzantine imperialism. The economic demands of aggregated settlement and centralized authority led to expanded orchard cultivation at the expense of natural woodlands. Even when these woodlands regenerated during periods of political or economic recession, they contained entirely new combinations of major tree taxa as a legacy of human intervention. Against the oscillations of urban civilization, an ever-present backdrop of mobile pastoralism became enhanced in the face of centralized authority and during its collapse.

Thus, the highly humanized landscape of the modern Eastern Mediterranean is a product of compound cultural influences and biotic responses. These influences began with Neolithic impacts that were most significant for their localized intensity and very early inception. Subsequent impacts during the middle and late Holocene assumed regional proportions and followed trajectories intimately linked with the rise and collapse of urbanized civilization. In overview, this brief synthesis reveals most clearly the extent to which the evolution of Holocene natural environments is inexorably linked to the development of human societies that exploit them.

ACKNOWLEDGMENTS

The analysis of paleobotanical remains from hyrax middens was supported in part by a W. F. Albright Fellowship awarded to Patricia Fall by the American Schools of Oriental Research. Middens were collected in the vicinity of Petra under permit from the Department of Antiquities of Jordan. The American Center of Oriental Research provided a jeep and research facilities in Amman. We offer our sincere thanks to these organizations for their generous assistance and hospitality. Additional thanks go to Suzanne Fish, Charles Redman, Alan Simmons, and David Steadman for their helpful commentary on preliminary versions of this chapter.

REFERENCES

Adams, R. M. (1978). Strategies of maximization, stability, and resilience in Mesopotamian society, settlement, and agriculture. *Proceedings of the American Philosophical Society* **122**, 329-335.

Adams, R. M. (1981). "Heartland of Cities. Surveys of Ancient Settlement and Land Use on the Central Floodplain of the Euphrates." University of Chicago, Chicago.

Banning, E., and Byrd, B. (1984). The architecture of Pre-Pottery Neolithic B 'Ain Ghazal, Jordan. *Bulletin of the American Schools of Oriental Research* **255**, 15-20.

Banning, E., and Byrd, B. (1987). Houses and the changing residential unit: domestic architecture at Pre-Pottery Neolithic B 'Ain Ghazal, Jordan. *Proceedings of the Prehistoric Society* **53**, 309-325.

Baruch, U. (1986). The Late Holocene vegetational history of Lake Kinneret (Sea of Galilee), Israel. *Paleorient* **12/2**, 37-48.

Baruch, U. (1990). Palynological evidence of human impact on the vegetation as recorded in Late Holocene lake sediments in Israel. *in* "Man's Role in the Shaping of the Eastern Mediterranean Landscape" (Bottema, S., Entjes-Nieborg, G., and Zeist, W. van, Eds.), pp. 283-293. A.A. Balkema, Rotterdam.

Bar-Yosef, O. (1986). The walls of Jericho: an alternative interpretation. *Current Anthropology* **27**, 157-162.

Behre, K.-E. (1981). The interpretation of anthropogenic indicators in pollen diagrams. *Pollen et Spores* **23**, 225-245.

Behre, K.-E. (1990). Some reflections on anthropogenic indicators and the record of prehistoric occupation phases in pollen diagrams from the Near East. *In* "Man's Role in the Shaping of the Eastern Mediterranean Landscape" (Bottema, S., Entjes-Nieborg, G., and Zeist, W. van, Eds.), pp. 219-230. A.A. Balkema, Rotterdam.

Borowski, O. (1987). "Agriculture in Iron Age Israel." Eisenbrauns, Winona Lake, IN.

Bottema, S. (1982[1985]). Palynological investigations in Greece with special reference to pollen as an indicator of human activity. *Palaeohistoria* **24**, 257-289.

Bottema, S., and Barkoudah, Y. (1979). Modern pollen precipitation in Syria and Lebanon and its relation to vegetation. *Pollen et Spores* **21**, 427-480.

Bottema, S., and Woldring, H. (1990). Anthropogenic indicators in the pollen record of the Eastern Mediterranean. *In* "Man's Role in the Shaping of the Eastern Mediterranean Landscape" (Bottema, S., Entjes-Nieborg, G., and Zeist, W. van, Eds.), pp. 231-264. A.A. Balkema, Rotterdam.

Broshi, M. (1978). Estimating the population of ancient Jerusalem. *Biblical Archaeology Review* **4**, 10-15.

Broshi, M. (1979). The population of Western Palestine in the Roman-Byzantine Period. *Bulletin of the American Schools of Oriental Research* **236**, 1-10.

Butzer, K. W. (1975). Patterns of environmental change in the Near East during late Pleistocene and early Holocene times. *In* "Problems in Prehistory: North Africa and the Levant" (Wendorf, F. and Marks, A. E., Eds.), pp. 398-410. Southern Methodist University Press, Dallas.

Chisholm, M. (1968). "Rural Settlement and Land Use." Aldine, Chicago.

Dever, W. G. (1980). New vistas on the EB IV ("MB I") horizon in Syria-Palestine. *Bulletin of the American Schools of Oriental Research* **237**, 35-64.

Dever, W. G. (1987). The Middle Bronze Age. The zenith of the urban Canaanite era. *Biblical Archaeologist* **50**, 149-177.

Dollfus, G., Kafafi Z., Rewerski, J., Vaillant, N., Coquengniot, E., Desse, J., and Neef, R. (1988). Abu Hamid, an early fourth millenium site in the Jordan Valley. *In* "The Prehistory of Jordan" (Garrard, A. N. and Gebel, H. G., Eds.), pp. 567-601. British Archaeological Reports International Series 396, Oxford.

Fall, P. L. (1990). Deforestation in southern Jordan: evidence from fossil hyrax middens. *In* "Man's Role in the Shaping of the Eastern Mediterranean Landscape" (Bottema, S., Entjes-Nieborg, G., and Zeist, W. van, Eds.), pp. 271-281. A.A. Balkema, Rotterdam.

Fall, P. L., Lindquist, C. A., and Falconer, S. E. (1990). Fossil hyrax middens from the Middle East: a record of paleovegetation and human disturbance. *In* "Packrat Middens: The Last 40,000 Years of Biotic Change" (Betancourt, J. L., Van Devender, T. R., and P. S. Martin, Eds.), pp. 408-427. University of Arizona Press, Tucson.

Felix, Y. (1982). The Jewish agriculture in the land of Israel in the period of the Misna and the Talmud. *In* "Eretz Israel from the Destruction of the Second Temple to the Muslim Conquest, Vol. I" (Baras, Z., Safrai, S., Tsafrir, Y, and Stern, M., Eds.), pp. 420-441 (in Hebrew).

Fritz, V. (1987). Conquest or settlement? The early Iron Age in Palestine. *Biblical Archaeologist* **50**, 84-100.

Gebel, H. G., Muheisen, M. S., and Nissen, H. J. (1988). Preliminary report on the first season of excavations at Basta. *In* "The Prehistory of Jordan. The State of Research in 1986." (Garrard, A. N. and Gebel, H. G., Eds.), pp. 101-134. British Archaeological Reports, International Series 396, Oxford.

Gonen, R. (1984). Urban Canaan in the Late Bronze Period. *Bulletin of the American Schools of Oriental Research* **253**, 61-73.

Harrison, D. L. (1968). "The Mammals of Arabia, Vol. II." Ernest Benn, London.

Hennessy, J. B. (1969). Preliminary report on a first season of excavations at Teilelat Ghassul. *Levant* **1**, 1-24.

Hole, F. (1974). Tepe Tula'i: an early campsite in Khuzistan, Iran. *Paleorient* **2**, 219-242.

Kenyon, K. (1960). "Excavations at Jericho, vol. 1." British School of Archaeology, Jerusalem.

Köhler-Rollefson, I., Gillespie, W., and Metzger, M. (1988). The fauna from Neolithic 'Ain Ghazal. *In* "The Prehistory of Jordan. The State of Research in 1986" (Garrard, A. N. and Gebel, H. G., Eds.), pp. 423-430. British Archaeological Reports, International Series 396, Oxford.

Köhler-Rollefson, I. and Rollefson, G. O. (1990). The impact of Neolithic subsistence strategies on the environment: the case of 'Ain Ghazal, Jordan. *In* "Man's Role in the Shaping of the Eastern Mediterranean Landscape" (Bottema, S., Entjes-Nieborg, G., and Zeist, W. van, Eds.), pp. 3-14. A.A. Balkema, Rotterdam.

Leonard, A. (1989). The Late Bronze Age. *Biblical Archaeologist* **52**, 4-39.

Mellaart, J. (1967). "Çatal Hüyük: A Neolithic Town in Anatolia." McGraw-Hill, New York.

Moore, A. M. T. (1982). Agricultural origins in the Near

East: a model for the 1980s. *World Archaeology* **14**, 224-235.

Neef, R. (1990). Introduction, development and environmental implications of olive culture: the evidence from Jordan. *In* "Man's Role in the Shaping of the Eastern Mediterranean Landscape" (Bottema, S., Entjes-Nieborg, G., and Zeist, W. van, Eds.), pp. 295-306. A.A. Balkema, Rotterdam.

Pearson, G. W. (1983). "The Development of High Precision ^{14}C Measurement and its Application to Archaeological Time Scale Problems." Unpublished PhD Thesis, Queen's University, Belfast.

Pires-Ferreira, J. W. (1975-77). Tepe Tula'i: faunal remains from an early campsite in Khuzistan, Iran. *Paleorient* **3**, 275-280.

Richard, S. (1987). The Early Bronze Age. The rise and collapse of urbanism. *Biblical Archaeologist* **50**, 22-43.

Roberts, N. (1990). Human-induced landscape change in south and southwest Turkey during the later Holocene. *In* "Man's Role in the Shaping of the Eastern Mediterranean Landscape" (Bottema, S., Entjes-Nieborg, G., and Zeist, W. van, Eds.), pp. 53-67. A.A. Balkema, Rotterdam.

Rollefson, G., and Köhler-Rollefson, I. (1989). The collapse of early Neolithic settlements in the Southern Levant. *In* "People and Culture in Change: Proceedings of the Second Symposium on Upper Paleolithic, Mesolithic, and Neolithic Populations of Europe and the Mediterranean Basin" (Hershkovitz, I., Ed.), pp. 59-72. British Archaeological Reports, International Series 508, Oxford.

Rollefson, G., and Köhler-Rollefson, I. (1992). Early Neolithic exploitation patterns in the Levant: cultural impact on the environment. *Population and Environment* **13**, 243-254.

Rollefson, G., and Simmons, A. (1986). The Neolithic village of 'Ain Ghazal, Jordan: preliminary report on the 1984 season. *Bulletin of the American Schools of Oriental Research Supplement* **24**, 147-164.

Rollefson, G., Simmons, A., and Kaffafi, Z. (1992). Neolithic cultures at 'Ain Ghazal, Jordan. *Journal of Field Archaeology* **19**, 443-470.

Shmida, A. (1980). Kermes oak in the land of Israel. *Israel Land and Nature* **6**, 9-16.

Stager, L. E. (1985). The firstfruits of civilization. *In* "Palestine in the Bronze and Iron Ages. Papers in Honor of Olga Tufnell" (Tubb, J. N., Ed.), pp. 172-188. University of London, Institute of Archaeology, Occasional Papers 11, London.

Zeist, W. van (1967). Late Quaternary vegetation history of western Iran. *Review of Palaeobotany and Palynology* **2**, 301-311.

Zeist, W. van, and Bottema, S. (1982). Vegetational history of the Eastern Mediterranean and the Near East during the last 20,000 years. *In* "Paleoclimates, Palaeoenvironments and Human Communities in the Eastern Mediterranean Region in Later Prehistory" (Bintliff, J. L. and Zeist, W. van, Eds.), pp. 277-321. British Archaeological Reports, International Series 133, Oxford.

Zeist, W. van, and Woldring, H. (1978). A postglacial pollen diagram from Lake Van in east Anatolia. *Review of Palaeobotany and Palynology* **26**, 249-276.

Zeist, W. van, Woldring, H., and Stapert, D. (1975). Late Quaternary vegetation and climate of southwestern Turkey. *Palaeohistoria* **17**, 53-143.

Zohary, M,. and Spiegel-Roy, P. (1975). Beginnings of fruit growing in the Old World. *Science* **187**, 319-327.

ENDNOTE

[1]Note: Based on his comparison with pollen zones in a core from Lake Kinneret, Baruch (1990) suggests that the radiocarbon dates at Ein-Gedi may be 2300-2400 years too old, thereby revising the basal age for the core to approximately 5600 yr B.P. He offers no reason why the radiocarbon ages should be inaccurate.

Chapter 8

Utilization of Sedimentological, Geochemical, and Palynological Information to Determine Paleo-Functions of Ancient Coal-Forming Wetlands

ELEANORA I. ROBBINS

Department of the Interior, U.S. Geological Survey, MS 956, Reston, Virginia 22092

INTRODUCTION

The determination of wetland function has become a major objective of wetland science (Adamus and Stockwell, 1983; Brinson, 1993b; Brinson *et al.*, 1993; Greeson *et al.*, 1979). These functions are the normal or characteristic processes and roles of wetland ecosystems. Wetlands form at the transition between land and water (Cowardin *et al.*, 1979), placing them in positions to be subjected to continual development pressure. This pressure has made it very important for wetland scientists to define terms precisely, to refine criteria to separate functions from human values, and to help lawmakers understand the natural functions.

Coal scientists have not needed to focus on the functions of coal-forming wetlands. Instead, environmental concerns are driving new studies of ash and sulfur content, trace metal composition, and properties in processing (Finkelman, 1993a, b; Pokorny, 1993). Coal is considered to be the result of deposition of plants and algae in ancient swamps; therefore, botanical composition is a major additional study topic (Stach *et al.*, 1982). In order to study detailed geochemical, sedimentological, and botanical composition, coal is collected from cores or columns cut through coal beds in outcrops and mine faces (Golightly and Simon, 1989). Such cuts reveal the location and composition of splits or partings, which are mineral layers between the coal beds.

Data on the geochemical, sedimentological, and botanical composition of coal can also be used to help analyze the paleo-functions of ancient coal swamps. Additional useful information about ancient functions and the role of the ancient wetlands in the paleo-landscape can be gleaned from the study of facies relations. Many studies have been published on the facies or the rocks of equivalent paleontological age that lay upstream and downstream of coal swamps (Diessel, 1992; McCabe and Parrish, 1992; Rahmani and Flores, 1984; Scott, 1987.)

The major objective of this paper is to analyze data from coal deposits and wetlands in order to determine whether such data can lead to better understanding of wetland and coal-swamp function. Examples from the literature

Late Quaternary Environments and Deep History: A Tribute to Paul S. Martin, edited by David W. Steadman and Jim I. Mead. The Mammoth Site of Hot Springs, South Dakota, Inc. Scientific Papers, Volume 3. Hot Springs, South Dakota, 1995.

will be used in this preliminary analysis. Furthermore, humans have made major modifications to the landscape by building roads and dams, clear-cutting forests, draining wetlands, polluting aquifers, channelizing rivers, ditching, allowing overgrazing, and introducing exotic species. The study of pristine ancient wetland ecosystems should lead to insights about natural processes that can only be partially observed in the highly modified modern wetlands.

DEFINITIONS

Wetlands are delineated on the basis of soils, plants, and hydrology. They are places in the landscape where the soils are moist or wet (hydric), where the plants are hydrophytic and therefore adapted to high water tables or flooding, and where the water table is at or near the surface (Brinson, 1991; Cowardin *et al.*, 1979; Sipple, 1988; Tiner, 1993).

Classification of wetland types has become an important topic in wetland science (Hammer, 1992; Mitsch and Gosselink, 1986). In the Cowardin *et al.* (1979) classification, wetlands are found in five major systems or environments: marine, estuarine, riverine, lacustrine, and palustrine. In the hydrogeomorphic classification, the major types are determined by their geomorphic setting (riverine, depressional, fringe of ocean or lake, extensive peatlands), their water source (precipitation, overland flow, ground-water discharge, sea water), and their hydrodynamics (vertical fluctuation, unidirectional or bidirectional horizontal flow) (Brinson, 1993a; Smith, 1993).

Many different terms are used for wetlands, depending on the country and its geologic history. Wetlands or mires can be defined by vegetation (for example, swamp or marsh), by chemistry and vegetation (for example, acid bog or alkaline fen), or by the absence or presence of peat (peatlands) (Gore, 1983; Hammer, 1992). Peatlands are the precursors to coal swamps. This paper uses "coal swamp" as a general term for all ancient peat-forming environments (McCabe and Shanley, 1992).

Peat can be composed of a mixture of many different organic and inorganic substances, including the remains of plants, algae, bacteria, and fungi, as well as remains of the herbivore and detritovore population of insects and worms. Peat is defined in the United States as an organic soil containing less than 25% inorganic components (see Andrejko *et al.*, 1983). In other countries, the inorganic content need only be less than 50%.

Coal is defined in the United States as a combustible sedimentary rock containing either less than 25% ash (ASTM, 1991) or less than 50% ash (Wood *et al.*, 1983). Ash is the inorganic component that remains after low or high temperature ashing, including the metals of enzymes, the cations and anions that participated in ion exchange reactions, mineral remains of organisms such as siliceous algae, chemical input from ground water and rain water, and mineral components blown or washed into the wetland. Ash can also be the mineral components added by through-flowing ground water after peat burial. To further confuse the terminology, it also may include volcanic ash that falls into coal swamps; in this form it is called tonstein (Stach *et al.*, 1982).

Depositional models for coal swamps are based on modern analogs. Coal scientists use a geomorphic approach, and models have been developed for coals deposited in back-barrier settings (Hobday and Horne, 1977), on river deltas (Belt *et al.*, 1984; Scott, 1989), within tectonic basins (Flores, 1986; Robbins, 1983; Robbins *et al.*, 1988), around lakes (Roehler and Martin, 1987), along rivers (Flores and Hanley, 1984), in estuarine settings (Breyer, 1987), and as domes on coastal plains (Cecil *et al.*, 1985; Shanley and McCabe, 1993).

TABLE 1. Selected functions of modern wetlands (adapted from Adamus and Stockwell, 1983; Brinson, 1993b; Carter, 1986; Gannon *et al.*, 1979; and Smith, 1993).

Habitat Creation
- Provide habitat for primary producers: plants and algae
- Provide habitat for consumers: invertebrates and vertebrates
- Maintain biological diversity/abundance

Water-Quantity Regulation
- Modify floodflow (storage and desynchronization)
- Discharge ground water to maintain stream and lake water levels (base flow maintenance)
- Recharge underlying aquifers

Water-Quality Regulation
- Filter toxic substances and contaminants from upstream or onshore sources
- Perform biogeochemical transformations and internal processing
- Export solutes, nutrients, and detritus to downstream or offshore waters/estuarine water balance

Control of Sediment Transport
- Trap sediments from upstream sources, thereby protecting downstream lakes, rivers, and estuaries
- Protect shorelines from offshore waves and storms (coastal erosion)

Climate Modification
- Modify atmospheric conditions
- Store and release carbon/produce greenhouse gases

Geologic Archive
- Serve as modern analogs
- Serve as archive repository of geologic history

WETLAND FUNCTIONS

Methodologies to assess functions of modern wetlands vary depending on the purpose of the analysis. The various Federal agencies that regulate wetlands in the United States have methodologies that primarily focus on quantifiable functions having results that can be enforced (Sifneos *et al.*, 1992). These functions include biotic processes, regulation of water quantity and quality, and control of sediment transport. Wetlands can also modify climate and storms and serve as models for the study of ancient functions. Most wetland functions result from a combination of biological, hydrologic, geochemical, sedimentological, and climatic processes. Table 1, listing selected functions of modern wetlands, will be used to analyze ancient wetland function. The analysis will focus on the underlying biological, chemical, and physical processes from the point of view of a coal geologist.

HABITAT CREATION

Provide habitat for primary producers: plants and algae

Wetlands are home to characteristic plants, such as bald cypress, sedges, and rushes (Crum, 1988; Mitsch and Gosselink, 1986). The fossil record shows that each geologic age has had its own characteristic plants and algae that grew in the swamps and produced coal. Information about the coal swamp plant community is sought in coal balls (permineralized structures that preserve plant parts), macrofossil assemblages (preserved in coal fields by floods), and in palynological reconstructions. Preservation is a haphazard process, however, and much is not known about the species composition of the earliest coal-forming plant communities.

In the Late Carboniferous, 30-m-tall tree species of lycopods dominated the forested and clastic swamps of Germany and Pennsylvania (Stach *et al.*, 1982; Wnuk, 1984; Wnuk and Pfeffercorn, 1987). Seed ferns and horsetails formed Carboniferous coal deposits in the Ruhr basin, whereas the conifer *Cordaites* formed coal deposits in the USSR (Stach *et al.*, 1982). During the Jurassic and Early Cretaceous, ginkgophytes, cycadophytes, and conifers dominated most swamps. Plant remains in Tertiary coal beds are better preserved than in older floras, so reconstructions of the coal-forming vegetation are more detailed. For example, Stach *et al.* (1982) provided a detailed reconstruction of two plant communities, a Nyssaceae-Taxodiaceae swamp and a Myricaceae-Cyrillaceae bush moor or fen, that formed the Miocene brown coal in the Lower Rhine embayment of Germany.

Palynomorphs provide additional information about coal-forming vegetation. Phillips and Peppers (1984) analyzed changes in coal-forming vegetation in the Late Carboniferous of Euroamerica. The reconstruction showed how the lycopsid flora was succeeded by one rich in gymnosperms. On the basis of megaspores, Bertram (1987) reconstructed the vegetation of a coal-forming ecosystem in Yorkshire, England, showing three phases of succession as the swamp progressed from lycospore-dominated submergent plants to densospore-dominated emergent plants.

Fossils of lipid-rich algae such as *Botryococcus* are reported by coal petrologists as the maceral alginite in boghead coals (Stach *et al.*, 1982). Calcareous algae have also been identified in a sub-bituminous Pliocene coal from Italy (Stach *et al.*, 1982).

Provide habitat for consumers: invertebrates and vertebrates

Wetlands are home to a large variety of animals (Mitchell *et al.*, 1993), some of which have left a fossil record in coal fields. In coal beds, whole fossils of invertebrates, amphibians, reptiles, and mammals are rare, primarily because of bone and shell dissolution. Typically, animal fossils are identified from channel sequences and lakes penecontemporaneous with coal beds (Arkle *et al.*, 1979; Eggleston, 1992). However, microcrustacean fecal pellets, crayfish remains, lamellibranch shells, sponge spicules, mussels, and vertebrate skeletons have been reported in coal beds of various ages from Germany, Greece, England, Italy, and North Carolina (Robbins, 1987; Stach *et al.*, 1982). Fish have been found in cannel and boghead coals and shaley interbeds that resulted from marine or lacustrine transgressions over coal swamps (Lomax, 1926; Stach *et al.*, 1982). At Baccinello, Italy, an almost complete skeleton of the primate *Oreopithecus* was found in a Pliocene coal bed.

Maintain biological diversity/abundance

Diversity and productivity in wetlands can be low or high depending on the wetland type. For example, vascular plants are highly diverse, as many as 100 species an acre, where water tables fluctuate such as in the wet prairies of Illinois (G.S. Wilhelm, personal communication, 1993), whereas vascular plant diversity is very low in stable, acid bogs (Crum, 1988). Above and below ground productivity is measured in wetlands and has even been estimated in coal beds (Phillips and Peppers, 1984). Unlike the lack of studies on most paleo-functions, paleo-diversity in coal swamps has been the focus of several studies. DiMichele *et al.* (1985), DiMichele and Hook (1992), and Wnuk (1984) reported on rich floras from Pennsylvanian coal deposits. DiMichele and Hook (1992) discussed such issues as floristic conservation, and temporal changes in diversity and structural complexity in Pennsylvanian and Permian coal beds. The fossil record shows that diversity may be increasing in coal-depositing plant communities with time, although this information must be analyzed carefully, because it is more likely that preservation is better in younger coals (Stach *et al.*, 1982).

WATER-QUANTITY REGULATION

Hydrologic processes active in past environments are difficult to analyze because water may not leave behind direct fossil evidence. Therefore, circumstantial evidence must be used to analyze the paleo-function of water-quantity regulation.

Modify floodflow (storage and desynchronization)

Flood and storm waters typically flow from uplands into lower lying wetlands. Wetlands then temporarily store the water, releasing it gradually, thereby desynchronizing the flow to even lower lying areas (Carter, 1986; Kadlec, 1993) . Large floods can carry large sediment loads (Hupp *et al.*, 1993) that can both disrupt and nourish the plants and animals living in the wetlands.

Large ancient floods would leave sedimentological evidence in the fossil record, along with the very finest sediment fractions that got deposited as storm and floodwaters abated. Therefore, even though there is probably no way to learn if ancient coal swamps served to desynchronize floods, it might be useful to explore sedimentological data to support such a paleo-function. In the Upper Freeport coal bed of Pennsylvania, Ruppert *et al.* (1991) found the highest amounts of detrital minerals in the vicinity of paleo-channels, showing the presence of ancient flood events in the Pennsylvanian. Thin splits in Carboniferous coal beds consisting of claystone or shale delineated in the Wilhelm coal bed of Germany (Chao *et al.*, 1983) could be evidence of flood waters that deposited fine-grained sediment during temporary storage.

Discharge ground water to maintain stream and lake water levels (base flow maintenance)

By storing water and releasing it slowly, some wetlands serve to reduce and others to maintain base flow within their local rivers (Carter, 1986; Phillips *et al.*, 1993). Although such a paleo-function can not be proved, it is possible to determine which streams and lakes were penecontemporaneous with coal beds. Coal beds are traced laterally to learn the nature of the associated facies. Fluvial sequences are common facies equivalents of coal fields (Rahmani and Flores, 1984). For example, Flores and Hanley (1984) reported on anastomosing rivers associated with the Cretaceous Powder River coal beds in Wyoming, and Rust *et al.* (1984) found similar associations with Pennsylvanian coal beds of the Cumberland Group in Nova Scotia. The association suggests that these ancient wetlands might have served to maintain base flow in the through-flowing paleo-river systems.

Recharge underlying aquifers

By storing water, some modern wetlands recharge underlying aquifers. For example, the Everglades is a source of recharge to the underlying Biscayne limestone aquifer that supplies southeastern Florida with drinking water (Fish, 1987; McPherson *et al.*, 1976). The karstification of this aquifer is thought to be aided by leaching from organic acids from the overlying wetland (T.M. Scott, personal communication, 1993). Some ancient coal swamps probably fulfilled this same recharge function, as indicated by a karstified limestone underlying a coal bed. An example is the Mulberry coal bed (Pennsylvanian) lying on the St. Joe (Mississippian) Limestone in Linn County, Kansas (Chestnut, 1993; Cobb, 1981).

WATER-QUALITY REGULATION

Removal of toxic substances and contaminants from upstream or onshore sources

Wetlands can trap or precipitate deleterious substances and metals and thus regulate water quality downstream (Brinson, 1993b). Removal processes include uptake by plants, microbial redox reactions, chemical and biochemical adsorption by peat, and sedimentation of particulate matter (Odum, 1979). These processes have been analyzed in wetland studies that have focused on the trapping of lead, uranium, selenium, and copper (Hupp *et al.*, 1993; Owen, 1992; Presser and Barnes, 1985; Presser *et al.*, 1990; Puckett *et al.*, 1993; Walton-Day *et al.*, 1990).

Over 90 elements have been found in coal beds, including environmentally deleterious elements such as As, Cd, Cr, Hg, Mn, Pb, Se, U, and Zn (Coleman *et al.*, in press; Gluskoter *et al.*, 1977; Tewalt *et al.*, 1993; Zubovic *et al.*, 1980). These trace elements have several pathways of entry into coal—one is being trapped during peat deposition, another is being trapped in buried peat and coal as these permeable beds later serve as aquifers.

Brownfield *et al.* (1991) found that chromium had naturally accumulated in a swamp from chromium-bearing dunites in the paleowatershed of a Tertiary coal bed from Washington. A postdepositional, epigenetic trapping function is documented by Ting (1983) who reported on silica concentrated in North Dakota lignite, which is now an important aquifer (Horak, 1983). Cadmium and zinc in Pennsylvanian coal beds in the Illinois basin are thought to be postdepositional components from brines carrying metal sulfides of Pb, Zn, and Cd that may have been mobilized by postdepositional tectonic compression (Cobb, 1981; Cobb *et al.*, 1980; Sverjensky and Garven, 1992).

Perform biogeochemical transformations and internal processing

Redox conditions within modern wetlands can be very complex. The presence of oxygenated, suboxic, and anoxic microenvironments leads to numerous microhabitats and diverse faunal and floral populations (Bandyopadhyany *et al.*, 1993; Odum, 1979; Robbins *et al.*, 1992). These organisms perform internal processing leading to biogeochemical transformations that play a major role in the filtration function of wetlands (Godfrey *et al.*, 1985; Hemond *et al.*, 1987; Mitsch and Gosselink, 1986).

Whereas it may prove impossible to reconstruct such complexity in the fossil record, it is possible to monitor the transformations of one element in coal fields because of its troublesome nature. Sulfur, in the form of pyrite, is often considered a deleterious element because of its role in the formation of acid mine drainage and acid rain. Coal may sometimes be analyzed for other major elements, but it always is analyzed for sulfur. Whatever the form of the original input of sulfur, whether it was reduced sulfide ion in anoxic ground water or oxidized sulfate ion from oxygenated marine or ground water, in oxygen-starved peat where iron is present, pyrite will become fixed (Casagrande, 1987).

Carboniferous coal beds from the Interior Coal Province have the highest content of pyritic-sulfur in United States (Davis, 1982). Internal processing reactions in the coal swamp probably were intense, although some of the sulfur probably was postdepositional. Williams and Keith (1963) showed that pyritic-sulfur values in some Appalachian coals were highest where marine rocks lay on top, suggesting that the sulfur was incorporated by the peat after the death of the wetland. Therefore, an unequivocal interpretation of internal processing reactions may not be possible in coal beds.

Export solutes, nutrients, and detritus to downstream or offshore waters/estuarine water balance

Wetlands exert a profound control on the quality of downstream or offshore waters (Mitsch and Gosselink, 1986). Outwelling of nutrients from coastal wetlands is an important process by which riverine and marine fisheries receive wetland nutrients (Odum, 1980). Furthermore, tides can carry some of the soluble and particulate fractions from coastal wetlands out to sea (de la Cruz, 1979; Haines, 1979). Exported detritus includes terrestrial carbon, particularly in the form of plant cells. The flow of water, solutes, nutrients, and detritus from wetlands to the offshore keeps estuaries as highly productive mixing zones between fresh water and sea water (Carter, 1986).

Although it probably is not possible to provide data from coal beds and their facies equivalents to prove paleo-regulation of water quality downstream from a coal swamp, the presence of terrestrial plant fossils in estuarine rocks may provide proxies for this paleo-function. Marine beds downstream from coal-depositing swamps would be the place to look for evidence. For example, the Mancos Shale downstream from coal of the John Henry Member of the Straight Cliffs Formation in the Kaiparowits field of Utah contains terrestrial debris and palynomorphs of Cretaceous wetland plants (Cross and Singh, 1975; Ryer, 1984). Holmes *et al.* (1992) found terrestrial isotope signatures in the $\partial^{13}C$ of marine rocks downstream from Cretaceous coal beds in the Frontier Formation of Wyoming.

CONTROL OF SEDIMENT TRANSPORT

Trap sediments from upstream sources, thereby protecting downstream lakes, rivers, and estuaries

By trapping sediment from onshore sources, wetlands serve to control the sediment input into downstream lakes, rivers, or estuaries such as Chesapeake Bay (Boto and Patrick, 1979; Hupp *et al.*, 1993; Puckett *et al.*, 1993). This function also serves to preserve habitat and maintain water quality (Mitsch and Gosselink, 1986). Sediment enters wetlands in a variety of ways. Rare large floods undoubtedly provide the most sediment, but the more frequent alluvial and colluvial inputs are also important (Hupp and Morris, 1990).

The evidence needed to support a sediment trapping paleo-function would consist of deposits of different sizes and geometric shapes. The deposits from large floods would tend to be coarse grained, and sediment would be deposited in a graded manner. Splits or partings of sediment between coal benches are the best sedimentological evidence for the paleo-sediment control function.

Studying thick Paleocene coal beds in the Tongue River Member of the Fort Union Formation in the Powder River basin of Wyoming and Montana, Ayers and Kaiser (1984) identified more than 30 splits that subdivide an otherwise thick sequence of coal. The splits were composed of sand and mud interpreted to be the result of transgressing deltas and interfingering distributary rivers. In a study of coal in two basins in India, Casshyap and Tewari (1984) identified sediment sources into the Permian coal swamp. Sandstone splits were considered to be the products of river channels and crevasse splays whereas carbonaceous shale splits were interpreted as deposits formed in the back swamp areas of meandering rivers. Studying mechanisms of split deposition in coal beds of the northern Appalachian Plateau, Ferm and Staub (1984) believed that flooding was the major cause of sediment input into coal swamps there. Li Sitian *et al.* (1984) found the major source of sediment into the late Mesozoic Fuxin coal basin in north-eastern China to be transgressing alluvial fans. All of these studies are good examples of the sediment trapping function of paleo-wetlands and of the variety of processes that transported sediment into ancient wetlands.

Protect shorelines from offshore waves and storms (coastal erosion)

Millions of dollars are spent annually to protect shorelines from expanding lakes, rising oceans, and major storm events such as hurricanes, cyclones, and tsunamis (Williams *et al.*, 1990). Coastal wetlands are a very important first-line defense against such natural disasters, particularly because they dissipate the energy (Gannon *et al.*, 1979). Coastal vegetation is typically adapted to periodic inundation; in fact, such transgressions are thought to be important to the nutrient budget and the health and maintenance of these coastal ecosystems (Mitsch and Gosselink, 1986).

This shoreline protection paleo-function of coal swamps can be analyzed as abrupt facies changes lateral to coal beds. Storm events would have caused abrupt facies changes or left pods of poorly sorted sediment along one edge of a coal field. Evidence for this paleo-function can be found in several studies using such criteria. Shanley and McCabe (1993) analyzed Upper Cretaceous coal beds in the Kaiparowits Plateau of Utah and found long-lived interfingering at the northeast part of the Plateau between coal of the John Henry Member of the Straight Cliffs Formation and the Bluegate Member of the Mancos Shale. This suggests that the John Henry coal swamp was able to protect the landward deposits, the Straight Cliffs Formation, for an estimated tens of thousands of years (P.J. McCabe, personal communication, 1993). Hower *et al.* (1987) found rip-up clasts of coal along one edge of the Herrin coal bed in Illinois and thereby identified the coastal edge of a domed coal swamp being protected from storm waves.

CLIMATE MODIFICATION

Modify atmospheric conditions

Gannon *et al.* (1979) showed that wetlands modify temperature and moisture content of the lower atmosphere, depending on areal extent, distribution, and depth of associated lakes. Wetlands also ameliorate freeze conditions, modify lake and sea breezes, and affect the behavior of tropical cyclones after landfall.

Such processes would leave no fossil record. However, DiMichele *et al.* (1985) and Moore (1987) have speculated that the regional climate of the great coal-forming periods, such as the Carboniferous and the Cretaceous, may have been greatly influenced by the coal swamp vegetation because climate and vegetation are intimately linked (Bonan *et al.*, 1992; Ziegler *et al.*, 1987).

Store and release carbon/produce greenhouse gases

Peat and coal deposits both represent large reservoirs of stored carbon that is not readily available to be converted into greenhouse gases without burning (Beamish and Vance, 1992). The amount of carbon stored in coal of all ages has been estimated at 31.5-45 x 10^{16} moles/C (Kasting and Walker, 1992). Modern-day use of such large reservoirs of fossil fuels leads the discussions on production of greenhouse gases (Emanual *et al.*, 1980). Most of the gases that increase the greenhouse affect (CH_4, CO, CO_2, light hydrocarbons, SO_2, DMS) are produced in wetlands (Cohen *et al.*, 1984).

Climate is also indirectly linked to carbon storage and CO_2 production. Haszeldine (1987) has suggested that the Earth's climate would have cooled as CO_2 became incorporated into the plants of the vast Carboniferous swamps. Water tables probably lowered as paleo-peat swamps switched from domal to planar morphologies in response to drying paleo-climates at the end of the Carboniferous (Cecil, 1990; Cecil *et al.*, 1985). Fires would have been more prevalent, and the release of CO_2 could have influenced atmospheric composition (Andreae, 1993).

GEOLOGIC ARCHIVE FUNCTIONS

Serve as modern analogs

Modern peatlands are being assessed for their use as analogs for low sulfur coal and depositional models. Neuzil *et al.* (in press) are studying Indonesian domed peat swamps as modern analogs for low-sulfur coal in the Appalachian coal field. Microtome sections of peat show that deposition of below-ground plant parts has been the dominant depositional mode of peat since the Paleozoic (Cohen and Spackman, 1972).

Serve as archive repository of geologic history

Important events in geologic history are discernable in studies of peat deposits and coal beds. Evolution of plants and environments, changing climates, catastrophic events, and variability in plant/animal interactions through time are all issues that have been addressed in the study of coal (DiMichele *et al.*, 1985; McCabe and Parrish, 1992; Rahmani and Flores, 1984; Scott, 1987; Stach *et al.*, 1982). Palynological analyses of peat have resulted in detailed knowledge of Quaternary vegetation and climatic changes (Bryant and Holloway, 1985; Shane and Cushing, 1991; Wright, 1983; Wright and Frey, 1965). Charcoal analyses have documented ancient catastrophic fires that affected coal swamps (Scott and Jones, 1991).

POSTDEPOSITIONAL FUNCTIONS

Coal itself has properties that act after it is buried. These are postdepositional functions. Coal has a passive tectonic function in that it can serve as graphite-like glide planes during faulting (Wood *et al.*, 1986) and perhaps even can lower earthquake intensity (Bauer, 1990). Lignite and coal may be important aquifers long after the wetland ceases to function and its peat is buried (Horak, 1983). Finally, buried coal can trap metals that are carried by formation waters that pass through it (Cobb, 1981). This is a very important postdepositional function of wetlands because those metals stay trapped in coal, rather than being discharged into the aquatic environment.

CONCLUSIONS

The functions of ancient wetlands can be discerned by studying the sedimentological, geochemical, and palynological aspects of coal beds and their associated sedimentary rocks. Information about the various functions of coal swamps is available in the literature even though most studies have not been performed specifically to test the functional aspect.

Paleo-habitat functions can be discerned in studies of changing coal-swamp vegetation and animal fossils. Paleo-water-quantity regulation is evident from the sediment record. Paleo-water quality is indicated by trace elements in coal and isotopic composition of plant tissues in estuarine rocks downstream from coal beds. Paleo-control of sediment transport is well documented as facies changes at the peripheries of coal fields. Paleo-climate modification cannot be directly analyzed, but speculation has been published on the climatic effect of extensive wetland habitats in earlier times. Finally, many aspects of geological history can be determined from detailed analysis of coal and peat.

This study has not discerned functions that were performed by ancient wetlands but not modern ones. There may have been times when ancient wetlands were more widespread and therefore had greater effect when performing their various functions, such as during the Carboniferous. There may have been places where ancient wetlands were extremely important landscape features, such as along the Cretaceous seaway of North America. This may contrast with the patchy distribution of wetlands today along the Atlantic seaway, which is also the location of human major population centers. The great thickness of coal and its widespread occurrence suggest that the paleofunctions of ancient coal swamps may have been more important in the past than today.

ACKNOWLEDGEMENTS

I would like to thank M.M. Brinson, V. Carter, G.G. Hollands, M.W. LaSalle, and R.D. Smith for sharing current ideas on wetland functions.

REFERENCES

Adamus, P.R., and Stockwell, L.T. (1983). "A Method for Wetland Functional Assessment" Vols. 1 and 2. U.S. Department of Transportation, Federal Highway Administration Reports FHWA-IP-82-23 and FHWA-IP-82-24.

Andreae, M.O. (1993). The influence of tropical biomass burning on climate and the atmospheric environment. *In* "Biogeochemistry of Global Change, Radiatively Active Trace Gases" (R.S. Oremland, Ed.), pp. 113-150, New York, Chapman and Hall, New York.

Andrejko, M.J., Fiene, F., and Cohen, A.D. (1983). Comparison of ashing techniques for determination of the inorganic content of peats. Testing of Peats and Organic Soils, American Society for Testing and Materials, ASTM STP 820, 5-20.

American Society for the Testing of Materials [ASTM]. (1991). "Annual Book of ASTM Standards, Sec. 5, Petroleum Products, Lubricants, and Fossil Fuels." Philadelphia, American Society for the Testing of Materials.

Arkle, T., Jr., Beissel, D.R., Larese, R.E., Nuhher, E.B., Patchen, D.G., Smosna, R.A., Gillespie, W.H., Lund, R. Norton, C.W., and Pfefferkorn, J.W. (1979). "The Mississippian and Pennsylvanian (Carboniferous) Systems in the United States—West Virginia and Maryland." Chapter D. *U.S. Geological Survey Professional Paper* **1110**, D1-D35.

Ayers, W.B., and Keiser, W.R. (1984). Lacustrine-interdeltaic coal in the Fort Union Formation (Palaeocene), Powder River Basin, Wyoming and Montana, U.S.A. In "Sedimentology of Coal and Coal-Bearing Sequences" (R.A. Rahmani and R.M. Flores, Eds.), pp. 61-84. *International Association of Sedimentologists, Special Publications* **7**.

Bandyopadhyay, B.K., Pezeshki, S.R., DeLaure, R.D., and Lindau, C.W. (1993). Influence of soil oxidation-reduction potential and salinity on nutrition, N-15 uptake, and growth of *Spartina patens*. *Wetlands* **13**, 10-15.

Bauer, E. (1990). Erfahrungen mit dem Bodenbewegungsmodell fuer das Saarkarbon. *Das Markscheidewesen* **97**, 455-459.

Beamish, B.B., and Vance, W.E. (1992). Greenhouse gas contributions from coal mining in Australia and New Zealand. *Journal of the Royal Society of New Zealand* **22**, 153-156.

Belt, E.S., Flores, R.M., Warwick, P.D., Conway, K.M., Johnson, K.R., and Waskowitz, R.S. (1984). Relationship of fluviodeltaic facies to coal deposition in the Lower Fort Union Formation (Palaeocene), south-western North Dakota. *In* "Sedimentology of Coal and Coal-Bearing Sequences" (R.A. Rahmani and R.M. Flores, Eds.), *International Association of Sedimentologists, Special Publication* **7**, 177-195.

Bertram, K.M. (1987). Lycopod succession in coals: an example from the Low Barnsley Seam (Westphalian B), Yorkshire, England. *In* "Coal and Coal Bearing Strata: Recent Advances" (A.C. Scott, Ed.), *Geological Society (London), Special Publication* **32**, 187-199.

Bonan, G.B., Pollard, D.D., and Thompson, S.L. (1992). Effects of boreal forest vegetation on global climate. *Nature (London)* **359**, 716-718.

Boto, K.G., and Patrick, W.H. (1979). Role of wetlands in the removal of suspended sediments. *In* "Wetland Functions and Values: the State of our Understanding" (P.E. Greeson, J.R. Clark, and J.E. Clark, Eds.), pp. 479-489. Proceedings of the National Symposium on Wetlands, American Water Resources Association, Minneapolis, Minnesota.

Breyer, J.A. (1987). A tidal origin for coarsening-up sequences above two Wilcox lignites in east Texas. *Journal of the Geological Society of London* **44**, 463-469.

Brinson, M.M. (1991). Landscape properties of pocosins and associated wetlands. *Wetlands*, Special Issue ***11***, 441-465.

Brinson, M.M. (1993a). "A Hydrogeomorphic Classification for Wetlands." Wetland Research Program Technical Report WRP-DE-4, U.S. Army Engineer Waterways Experiment Station, Vicksburg, Mississippi.

Brinson, M.M. (1993b). Changes in the functioning of wetlands along environmental gradients. *Wetlands*, Special Issue **13**, 65-74.

Brinson, M.M., Krcuzynski, W., Lee, L.C., Nutter, W.L., Smith, R.D., and Whigham, D.F. (1993). Developing an approach for assessing the functions of wetlands. *In* "Wetlands of the World. Biogeochemistry, Ecological Engineering, Modelling and Management" (W.J. Mitsch and R.E. Turner, Eds.), Elsevier, Amsterdam.

Brownfield, M.E., Affolter, R.H., and Stricker, G.D. (1991). High chrome contents in Tertiary coals, northwestern Washington—a key to their depositional history. *Geological Society of America, Abstracts with Programs* **23**, 144.

Bryant, V.M., Jr., and Holloway, R.G., Eds. (1985). "Pollen Records of Late-Quaternary North American Sediments." American Association of Stratigraphic Palynologists Foundation. Austin, Texas.

Carter, V. (1986). An overview of the hydrologic concerns related to wetlands in the United States. *Canadian Journal of Botany* **64**, 364-374.

Casagrande, D.J. (1987). Sulphur in peat and coal. In "Coal and Coal-Bearing Strata: Recent Advances" (A.C. Scott, Ed.), *Geological Society (London), Special Publication* **32**, 87-105.

Casshyap, S.M., and Tewari, R.C. (1984). Fluvial models of the Lower Permian coal measures of Son-Mahanadi and Koel-Damodar Valley basins, India. *In* "Sedimentology of Coal and Coal-Bearing Sequences" (R.A. Rahmani and R.M. Flores, Eds.), *International Association of Sedimentologists, Special Publication* **7**, 121-147.

Cecil, C.B. (1990). Paleoclimate controls on stratigraphic repetition of chemical siliciclastic rocks. *Geology* **18**, 533-536.

Cecil, C.B., Stanton, R.W., Neuzil, S.G., Dulong, F.T., Ruppert, L.F., and Pierce, B.S. (1985). Paleoclimate controls on Late Paleozoic sedimentation and peat formation in the central Appalachian basin (U.S.A.). *International Journal of Coal Geology* **5**, 195-230.

Chao, E.C.T., Minkin, J.A., and Thompson, C.L. (1983). "Recommended Procedures and Methodology of Coal Description." *U.S. Geological Survey Circular* **894**.

Chestnut, D.R., Jr. (1993). "Geologic Highway Cross Sections: Interstate Highway 64, Farmers to Catlettsburg, Kentucky." *Kentucky Geological Survey, Map and Chart Series* 4, Series 11.

Cobb, J.C. (1981). "Geology and Geochemistry of Sphalerite in Coal." Illinois State Geological Survey, Final Report, Grant No. 14-08-0001-G-496.

Cobb, J.C., Steele, J.D., Treworgy, C.G., and Ashby, J.F. (1980). The abundance of zinc and cadmium in sphalerite-bearing coals in Illinois. *Illinois Mineralogist Notes* **74**, 1-28.

Cohen, A.D., and Spackman, W. (1972). Methods in peat petrology and their application to reconstruction of paleoenvironments. *Geological Society of America, Bulletin* **83**, 129-142.

Cohen, A.D., Casagrande, D.J., Andrejko, M.J., and Best, G.R., Eds. (1984). "The Okefenokee Swamp: its Natural History, Geology, and Geochemistry." Wetland Surveys, Los Alamos.

Coleman, L., Bragg, L.J., and Finkelman, R.B. (in press). Distribution and mode of occurrence of selenium in US coals. *Environmental Geochemistry and Health.*

Cowardin, L.M., Carter, V., Golet, F.C., and LaRoe, E.T. (1979). "Classification of Wetlands and Deepwater Habitats of the United States." U.S. Fish and Wildlife Service, Biological Services Program, Report FWS/OBS-79/31.

Cross, A.T., and Singh, H.P. (1975). Palynologic indicators of paleoenvironments of Upper Cretaceous coals. *Geological Society of America, Abstracts with Programs* **7**, 1042.

Crum, H.A. (1988). "A Focus on Peatlands and Peat Mosses." University of Michigan Press, Ann Arbor.

Davis, A. (1982). Sulfur in coal. *(Pennsylvania State University) Earth and Mineral Sciences* **51**, 18-21.

de la Cruz, A.A. (1979). Production and transport of detritus in wetlands. *In* "Wetland Functions and Values: the State of our Understanding" (P.E. Greeson, J.R. Clark, and J.E. Clark, Eds.), pp. 162-174. Proceedings of the National Symposium on Wetlands, American Water Resources Association, Minneapolis, Minnesota.

Diessel, K. (1992). "Coal-bearing Depositional Systems." Springer-Verlag, New York.

DiMichele, W.A., and Hook, R.W. (1992). Paleozoic terrestrial ecosystems, Chapter 5. In "Terrestrial Ecosystems Through Time, Evolutionary Paleoecology of Terrestrial Plants and Animals" (A.K.

Behrensmeyer, J.S. Damuth, W.A. DiMichele, R. Potts, H.D. Sues, and S.L. Wing, Eds.), pp. 202-235. University of Chicago Press, Chicago.

DiMichele, W.A., Phillips, T.L., and Peppers, R.A. (1985). The influence of climate and depositional environment on the distribution and evolution of Pennsylvanian coal-swamp plants. *In* "Geological Factors and the Evolution of Plants" (B.H. Tiffney, Ed.), pp. 223-256. Yale University Press, New Haven.

Eggleston, J.R. (1992). Sedimentology of the Upper Pennsylvania Redstone Limestone, Northern Appalachian basin. *In* "Paleoclimate Controls on Carboniferous Sedimentation and Cyclic Stratigraphy in the Appalachian Basin" (C.B. Cecil, C.F. Eble, N. Fedorko, B.M. Blake, and W.C. Grady, Eds.), *U.S. Geological Survey Open-File Report* **92-546**, 120-128.

Emanuel, W.R., Olson, J.S., and Killough, G.G. (1980). The expanded use of fossil fuels by the U.S. and the global carbon dioxide problem. *Journal of Environmental Management* **10**, 37-49.

Ferm, J.C., and Staub, J.R. (1984). Depositional control of minable coal bodies. *In* "Sedimentology of Coal and Coal-Bearing Sequences" (R.A. Rahmani and R.M. Flores, Eds.). *International Association of Sedimentologists, Special Publication* **7**, 275-289.

Finkelman, R.B. (1993a). Air toxics in coal: abundance, distribution, modes of occurrence, and textural relations. *In* "Coal-Energy and the Environment" (S.-H. Chiang, Ed.), pp. 801-804. Proceedings of the 10th Annual International Pittsburgh Coal Conference. Pittsburgh, Pennsylvania.

Finkelman, R.B. (1993b). Haxardous trace elements in coal: can we rely on modes of occurrence information to predict their removal?. *In* "Coal-Energy and Environment" (S.-H. Chiang, Ed.), pp. 8311-313. Proceedings of the 10th Annual International Pittsburgh Coal Conference. Pittsburgh, Pennsylvania.

Fish, J.E. (1987). "Hydrogeology, Aquifer Characteristics, and Ground Water Flow of the Surficial Aquifer System, Broward County, Florida." *U.S. Geological Survey Water Resources Investigations Report* **87-4034**.

Flores, R.M. (1986). Styles of coal deposition in Tertiary alluvial deposits, Powder River basin, Montana and Wyoming. In "The Paleoenvironmental and Tectonic Controls in Coal-Forming Basins in the United States" (P.C. Lyons and C.L. Rice, Eds.), *Geological Society of America, Special Paper* **210**, 79-104.

Flores, R.M., and Hanley, J.H. (1984). Anastomosed and associated coal-bearing fluvial deposits: Upper Tongue Member, Palaeocene Fort Union Formation, northern Powder River Basin, Wyoming, U.S.A. *In* "Sedimentology of Coal and Coal-Bearing Sequences" (R.A. Rahmani and R.M. Flores, Eds.), *International Association of Sedimentologists, Special Publication* **7**, 85-103.

Gannon, P.T., Sr., Bartholic, J.F., and Bill, R.G., Jr. (1979). Climatic and meteorological effects of wetlands. *In* "Wetland Functions and Values: the State of our Understanding" (P.E. Greeson, J.R. Clark, and J.E. Clark, Eds.), pp. 576-588. Proceedings of the National Symposium on Wetlands, American Water Resources Association, Minneapolis, Minnesota.

Gluskoter, H.J., Ruch, R.R., Miller, W.G., Cahill, R.A., Greher, G.B., and Kuhn, J.K. (1977). Trace elements in coal: occurrence and distribution. *Illinois State Geological Survey, Circular* **499**, 1-154.

Godfrey, P.J., Kaynor, E.R., Pelczarski, S., and Benforado, J., Eds. (1985). "Ecological Considerations in Wetlands Treatment of Municipal Wastewaters." Van Nostrand Reinhold, New York.

Golightly, D.W., and Simon, F.O. (1989). "Methods for Sampling and Inorganic Analysis of Coal." *U.S. Geological Survey Bulletin* **1823**.

Gore, A.J.P., Ed. (1983). "Mires: Swamp, Bog, Fen and Moor: Ecosystems of the World." Vol. 4A, General Studies. Elsevier, New York.

Greeson, P.E., Clark, J.R., and Clark, J.E., Eds. (1979). "Wetland Functions and Values: the State of our Understanding." Proceedings of the National Symposium on Wetlands, American Water Resources Association, Minneapolis, Minnesota.

Haines, E.B. (1979). Interactions between Georgia salt marshes and coastal waters; a changing paradigm. *In* "Ecological Processes in Coastal and Marine Systems" (R.J. Livingston, Ed.), *Marine Sciences* **10**, 35-46.

Hammer, D.A. (1992). "Creating Freshwater Wetlands." Lewis Publishers, London.

Haszeldine, R.S. (1987). Coal reviewed; depositional controls, modern analogues and ancient climates. *Geological Society (London), Special Publication* **41**, 289-308.

Hemond, H.F., Army, T.P., Nuttle, W.K., and Chen, D.G. (1987). Element cycling in wetlands: interactions with physical mass transport. *In* "Sources and Fates of Aquatic Pollutants" (R.A. Hites and S.J. Eisenreich, Eds.), *Advances in Chemistry, Series* **216**, 519-537.

Hobday, D.K., and Horne, J.C. (1977). Tidally influenced barrier island and estuarine sedimentation in the Upper Carboniferous of southern West Virginia. *Sedimentology* **18**, 97-122.

Holmes, C.W., M'Gonigle, J.W., and Dalrymple, G.B.

(1992). Delta (13)C of terrestrial and marine organic matter in Cretaceous rocks of the Western Interior of North America. *U.S. Geological Survey Circular* **1074**, 1-36.

Horak, W.F. (1983). "Hydrology of the Wibaux-Beach Lignite Deposit Area, Eastern Montana and Western North Dakota." *U.S. Geological Survey Water Resources Investigation WRI* **83-4157**, 1-89.

Hower, J.C., Trinkle, E.J., Graese, A.M., and Neuder, G.L. (1987). Ragged edge of the Herrin (No. 11) coal, western Kentucky. *International Journal of Coal Geology* **7**, 1-20.

Hupp, C.R., and Morris, E.E. (1990). A dendrogeomorphic approach to measurement of sedimentation in a forested wetland, Black Swamp, Arkansas. *Wetlands* **10**, 107-124.

Hupp, C.R., Woodside, M.D., and Yanosky, T.M. (1993). Sediment and trace element trapping in a forested wetland, Chickahominy River, Virginia. *Wetlands* **13**, Special Issue, 95-104.

Kadlec, J.A. (1993). Effect of depth of flooding on summer water budgets for small diked marshes. *Wetlands* **13**, 1-9.

Kasting, J.F., and Walker, J.C.G. (1992). The geochemical carbon cycle. *In* "Global Warming: Physics and Facts" (B.G. Levi, D.W. Hafemeister, and R.A. Scribner, Eds.), pp. 175-200. American Institute of Physics, New York.

Li Sitian, Li Baofang, Yang Shigong, Huang Jiafu, and Li Zhen (1984). Sedimentation and tectonic evolution of Late Mesozoic faulted coal basins in northwestern China. *In* "Sedimentology of Coal and Coal-Bearing Sequences" (R.A. Rahmani and R.M. Flores, Eds.), *International Association of Sedimentologists, Special Publication* **7**, 387-406.

Lomax, J. (1926). A fossil fish spine. *Colliery Guardian* **131**, 961.

McCabe, P.J., and Parrish, J.T., (Eds.) (1992). "Controls on the Distribution and Quality of Cretaceous Coals." *Geological Society of America, Special Paper* **267**, 1-407.

McCabe, P.J., and Shanley, K.W. (1992). Organic control on shoreface stacking patterns: Bogged down in the mire. *Geology* **20**, 741-744.

McPherson, B.F., Hendrix, G.Y., Klein, H., and Tyus, H.M. (1976). "The environment of South Florida." *U.S. Geological Survey Professional Paper* **1011**, 1-81.

Mitchell, J.C., Erdle, S.Y., and Pagels, J.F. (1993). Evaluation of capture techniques for amphibian, reptile, and small mammal communities in saturated forested wetlands. *Wetlands* **13**, Special Issue, 130-136.

Mitsch, W.J., and Gosselink, J.G. (1986). "Wetlands." Van Nostrand Reinhold, New York.

Moore, P.D. (1987). Ecological and hydrological aspects of peat formation. *In* "Coal and Coal-Bearing Strata: Recent Advances" (A.C. Scott, Ed.), *Geological Society (London), Special Publication* **32**, 7-15.

Neuzil, S.G., Supardi, Cecil, C.G., Kane, J.S., and Seodjono, K. (in press). Inorganic geochemistry of domed peat in Indonesia and its implication for the origin of mineral matter in coal. *Geological Society of America, Special Paper* **286**.

Odum, E.P. (1979). The value of wetlands: a hierarchical approach. *In* "Wetland Functions and Values: the State of our Understanding" (P.E. Greeson, J.R. Clark, and J.E. Clark, Eds.), pp. 16-25. Proceedings of the National Symposium on Wetlands, American Water Resources Association, Minneapolis, Minnesota.

Odum, E.P. (1980). The status of three ecosystem-level hypotheses regarding salt marsh estuaries: Tidal subsidy, outwelling, and detritus-based food chains. *In* "Estuarine Perspectives" (V.S. Kennedy, Ed.), pp. 485-495. Academic Press, New York.

Owen, D.E. (1992). "Uranium and Other Elements in Colorado Rocky Mountain Wetlands: a Reconnaissance Study." *U.S. Geological Survey Bulletin* **1992**, 1-33.

Phillips, P.J., Denver, J.M., Shedlock, R.J., and Hamilton, P.A. (1993). Effect of forested wetlands on nitrate concentration in ground water and surface water on the Delmarva Peninsula. *Wetlands* **13**, Special Issue, 75-83.

Phillips, T.L., and Peppers, R.A. (1984). Changing patterns of Pennsylvanian coal swamp vegetation and implications of climatic change on coal occurrence. *International Journal of Coal Geology* **3**, 205-255.

Pokorny, G. (1993). Going green. *Coal Voice* **16**, 12-15.

Presser, T.S., and Barnes, I. (1985). "Dissolved Constituents Including Selenium in Waters in the Vicinity of Kesterson National Wildlife Refuge and the West Grassland, Fresno and Merced Counties, California." *U.S. Geological Survey Water Resources Investigation WRI* **85-4220**, 1-73.

Presser, T.S., Swain, W.C., Tidball, R.R., and Severson, R.C. (1990). "Geologic Sources, Mobilization, and Transport of Selenium from the California Coast Ranges to the Western San Joaquin Valley; a Reconnaissance Study." *U.S. Geological Survey Water Resources Investigation WRI* **90-4070**, 1-66.

Puckett, L.J., Woodside, M.D., Libby, B., and Schening, M.R. (1993). Sinks for trace metals, nutrients, and sediments in wetlands of the Chickahominy River near Richmond, Virginia. *Wetlands* **13**, Special Issue, 105-114.

Rahmani, R.A., and Flores, R.M., Eds. (1984). "Sedimentology of Coal and Coal-Bearing Sequences." *International Association of Sedimentologists, Special Publication* **7**.

Robbins, E.I. (1983). Accumulation of fossil fuels and metallic minerals in active and ancient rift lakes. *Tectonophysics* **94**, 633-658.

Robbins, E.I. (1987). Paleoecology of the Niland tongue. *In* "Geological Investigations of the Vermillion Creek Coal Bed in the Eocene Niland Tongue of the Wasatch Formation, Sweetwater County, Wyoming" (H.W. Roehler and P.L. Martin, Eds.), Chapter E. *U.S. Geological Survey Professional Paper* **1314**, 74-102.

Robbins, E.I., D'Agostino, J.P., Ostwald, J., Fanning, D.S., Carter, V., and Van Hoven, R.L. (1992). Manganese nodules and microbial oxidation of manganese in the Huntley Meadows wetlands, Virginia, USA. In "Biomineralization Processes, Iron and Manganese" (H.C.W. Skinner and R.W. Fitzpatrick, Eds.), *Catena Supplement* **21**, 179-202.

Robbins, E.I., Wilkes, G.P., and Textoris, D.A. (1988). Coal deposits of the Newark rift system. *In* "Triassic-Jurassic Rifting, Continental Breakup and the Origin of the Atlantic Ocean and Passive Margins" (Warren Manspeizer, Ed.), Part B, Chapter 27, pp. 649-682. Elsevier, New York.

Roehler, H.W., and Martin, P.L., Eds. (1987). "Geological Investigations of the Vermillion Creek Coal Bed in the Eocene Niland Tongue of the Wasatch Formation, Sweetwater County, Wyoming." *U.S. Geological Survey Professional Paper* **1314A-L**.

Ruppert, L.F., Stanton, R.W., Cecil, C.B., Eble, C.F., and Dulong, F.T. (1991). Effects of detrital influx in the Pennsylvanian Upper Freeport peat swamp. *International Journal of Coal Geology* **17**, 95-116.

Rust, B.R., Gibling, M.R., and Legun, A.S. (1984). Coal deposition in an anastomosing-fluvial system: the Pennsylvanian Cumberland Group south of Joggins, Nova Scotia, Canada. *In* "Sedimentology of Coal and Coal-Bearing Sequences" (R.A. Rahmani and R.M. Flores, Eds.), *International Association of Sedimentologists, Special Publication* **7**, 105-120.

Ryer, T.A. (1984). Transgressive-regressive cycles and the occurrence of coal in some Upper Cretaceous strata of Utah, U.S.A. In "Sedimentology of Coal and Coal-Bearing Sequences" (R.A. Rahmani and R.M. Flores, Eds.), *International Association of Sedimentologists, Special Publication* **7**, 217-227.

Scott, A.C., Ed. (1987). "Coal and Coal-Bearing Strata: Recent Advances." *Geological Society (London), Special Publication* **32**.

Scott, A.C. (1989). Deltaic coals: an ecological and palaeobotanical perspective. *In* "Deltas, Sites and Traps for Fossil Fuels" (M.K.G. Whateley and K.T. Pickering, Eds.), pp. 309-316. Blackwell Scientific, Boston.

Scott, A.C., and Jones, T.P. (1991). Fossil charcoal; a plant-fossil record preserved by fire. *Geology Today* **7**, 214-216.

Shane, L.C.K., and Cushing, E.J., Eds. (1991). "Quaternary Landscapes." University of Minnesota Press, Minneapolis.

Shanley, K.W., and McCabe, P.J. (1993). Alluvial architecture in a sequence stratigraphic framework: a case history from the Upper Cretaceous of southern Utah, USA. *International Association of Sedimentologists, Special Publication* **15**, 21-56.

Sifneos, J.C., Cake, E.W., Jr., and Kentula, M.E. (1992). Effects of Section 404 permitting on freshwater wetlands in Louisiana, Alabama, and Mississippi. *Wetlands* **12**, 28-36.

Sipple, W.S. (1988). "Wetland Identification and Delineation Manual." Vols. 1 and 2. U.S. Environmental Protection Agency, Office of Wetlands Protection.

Smith, R.D. (1993). A conceptual framework for assessing the functions and values of wetlands. Wetland Research Program Technical Report WRP-DE-3, U.S. Army Engineer Waterways Experiment Station, Vicksburg, Mississippi.

Stach, E., Mackowsky, M.-Th., Teichmueller, M., Taylor, G.H., Chandra, D., and Teichmueller, R. (1982). "Stach's Textbook of Coal Petrology." 3rd Edition. Gebrueder Borntraeger, Berlin.

Sverjensky, D.A., and Garven, G. (1992). Geochemistry; tracing great fluid migrations. *Nature (London)* **356**, 481-482.

Tewalt, S.J., Oman, C.L., Bragg, L.J., and Finkelman, R.B. (1993). Distribution of mercury in Pennsylvania bituminous coal: a database perspective. *In* "Coal-Energy and the Environment" (S.-H. Chiang, Ed.), pp. 1083-1087. Proceedings of the 10th Annual Pittsburgh Coal Conference.

Tiner, R.W. (1993). The primary indicators method—a practical approach to wetland recognition and delineation in the United States. *Wetlands* **13**, 50-64.

Ting, F.T.C. (1983). Petrified Paleocene peat of Fort Union Group of North Dakota. *In* "Mineral Matter in Peat, its Occurrence, Form, and Distribution" (R. Raymond, Jr., and M.J. Andrejko, Eds.), pp. 209-213. Los Alamos National Laboratory Report LA-9907-OBES.

Walton-Day, K., Filipek, L.H., and Papp, C.S.E. (1990). Mechanisms controlling Cu, Fe, Mn, and Co profiles in part of the Wilson Creek fen, northeastern Minnesota. *Geochimica et Cosmochimica Acta* **54**, 2933-2946.

Williams, E.G., and Keith, M.L. (1963). Relationship between sulfur in coals and occurrence of marine roof beds. *Economic Geology* **58**, 720-729.

Williams, S.J., Dodd, K., and Gohn, K.K. (1990). "Coasts in Crisis." *U.S. Geological Survey Circular* **1075**.

Wnuk, C. (1984). "The Paleoecology and Growth of Arborescent Lycopods from an in situ Swamp Forest of Middle Pennsylvanian Age (Bernice Basin, Sullivan County, Pennsylvania)." Unpublished Ph.D. dissertation, University of Pennsylvania, Philadelphia.

Wnuk, Christopher, and Pfeffercorn, H.W. (1987). A Pennsylvanian-age terrestrial storm deposit: using plant fossils to characterize the history and process of sediment accumulation. *Journal of Sedimentary Petrology* **57**, 212-221.

Wood, G.H., Jr., Kehn, T.M., Carter, M.D., and Culbertson, W.C. (1983). "Coal Resource Classification System of the U.S. Geological Survey." *U.S. Geological Survey Circular* **891**.

Wood, G.H., Jr., Kehn, T.M., and Eggleston, J.R. (1986). Depositional and structural history of the Pennsylvania anthracite region. *In* "Paleoenvironmental and Tectonic Controls in Coal-forming Basins of the United States" (P.C. Lyons and C.L. Rice, Eds.), *Geological Society of America, Special Paper* **210**, 31-47.

Wright, H.E., Jr., Ed. (1983). "Late Quaternary Environments of the United States," Vols. 1 and 2. University of Minnesota Press, Minneapolis.

Wright, H.E., Jr., and Frey, D.G., Eds. (1965). "The Quaternary of the United States," Princeton University Press, Princeton.

Zeigler, A.M., Raymond, A.L., Gierlowski, T.C., Horrell, M.A. Rowley, D.B., and Lottes, A.L. (1987). Coal, climate and terrestrial productivity: the present and early Cretaceous compared. *In* "Coal and Coal-Bearing Strata, Recent Advances" (A.C. Scott, Ed.), *Geological Society (London)*, Special Publication **32**, 25-49.

Zubovic, Peter, Oman, C.L., Bragg, L.J., Coleman, S.L., Rega, N.H., Lemaster, M.E., Rose, H.F., and Golightly, D.W. (1980). "Chemical Analysis of 659 Coal Samples from the Eastern United States." *U.S. Geological Survey Open-File Report* **80-2003**.

Chapter 9

Reflections on the Development of the California Pollen Record

DAVID P. ADAM

Department of the Interior, U.S. Geological Survey, 345 Middlefield Road, Menlo Park, California 94025

INTRODUCTION

California offers tremendous variety in its landscapes and biomes, and the floral history of the state is of considerable significance for our understanding of both past and future global climatic changes. Numerous Quaternary geological processes have produced closed depressions suitable for the preservation of pollen records, including glaciation, volcanism, landsliding, faulting, tectonic warping, and deposition of alluvial dams. The combination of floral diversity and a wide range of depositional environments thus makes California a particularly attractive target for pollen studies. Despite the 'easy pickings,' however, the palynological community has been slow to respond to the opportunities.

This paper presents an overview of the development of the California pollen record to date, as well as some observations concerning future work. The view presented is a decidedly personal one in which I emphasize certain events, mostly during the past 30 years, that mark significant expansions in the practice of palynology in California. I have chosen my examples from work with which I was usually involved in some way, and which can be related, directly or indirectly, to the influence of Paul Martin. I have thus passed by many other worthwhile contributions; no slight is intended. A broader survey of California pollen work through about 1984 can be found in Adam (1985).

EARLY WORK

Palynology was first introduced as a tool for Quaternary studies in Sweden (von Post, 1916). Twenty years later, the first pollen study in western North America (Osvald, 1936) was published. The first pollen studies in California appeared soon thereafter (Hansen, 1942) as part of the amazing collaboration led by Luther Cressman, who formed an interdisciplinary team for work on northwestern Great Basin archeological sites long before the value of a team approach was widely appreciated (Cressman, 1942). These first California studies were associated with archeological sites

Late Quaternary Environments and Deep History: A Tribute to Paul S. Martin, edited by David W. Steadman and Jim I. Mead. The Mammoth Site of Hot Springs, South Dakota, Inc. Scientific Papers, Volume 3. Hot Springs, South Dakota, 1995.

exposed by the draining of Lower Klamath Lake. The shallow samples, from depths of up to 3.3 meters, barely scratched the surface.

For the next two decades, California palynology was virtually ignored. Henry Hansen looked at a few more samples (Frink and Kues, 1954), but concentrated his efforts in the Pacific Northwest. A brief report by Roosma (1958) described a pollen record from Searles Lake; a more detailed paper was promised but never appeared. Calvin Heusser (1960), who included three coastal sites in northern California as part of his extensive study of the Northwest Coast, conducted the most detailed work during this period.

Application of palynology to Early Pleistocene sediments of the Chagoopa Plateau, in the southern Sierra Nevada (Axelrod and Ting, 1961), produced equivocal results. It is unclear from the paper how many samples were studied and how those samples related to the stratigraphy of the sites. The precision claimed for taxonomy of the pollen grains seems unrealistic (nearly all grains were identified to the species level). The paper was unsuccessful in treating pollen grains as though they were fossil leaves, and Axelrod has been unenthusiastic about palynology ever since. His tremendous influence on the practice of paleobotany in the western U.S. may in part account for the strange lack of interest in the long-term California pollen record by the state's many universities. The Chagoopa Plateau sites should be reexamined to see if more detailed stratigraphic work is warranted.

INITIATION OF SYSTEMATIC WORK

Until 1960, then, California palynology was essentially either nonexistent or a peripheral element of studies to the north. The first continuing studies that actually focused on California began in 1963, when Paul Martin provided the means and encouragement for the present author to initiate work in the Sierra Nevada, which offered a welcome respite from the southern Arizona summer climate.

In order to interpret fossil pollen records from the Sierra, the first step was to collect transects of modern pollen surface samples across the range. Two transects were collected, one across Tioga Pass and one across Carson and Monitor Passes (Adam, 1964, 1967). These two transects established an elevational pollen zonation confirmed by later work (Davis *et al.*, 1985). The initial stratigraphic studies in the Sierra Nevada included three short sections from Yosemite National Park (a valley fill, an archeological site, and a spring mound) and a 4.4-m core from the Tahoe Basin. The archeological site, at Crane Flat, yielded the first pollen diagram from the Sierra (Fig. 1), and the results were rather discouraging: only one pollen type showed much systematic variability, and that type was an unknown! (We later identified it as *Arceuthobium*, a dwarf mistletoe). The other Yosemite sites proved more interesting, and showed much more variation through time. The diagram for the valley fill site, Hodgdon Ranch (Fig. 2), showed a major increase in TCT pollen at the expense of pine pollen from the bottom of the section to the middle, and then a reversal of that trend from the middle to the top; this was interpreted as a cool-warm-cool cycle. The Hodgdon Ranch record has turned out to be significant in interpreting some much longer records (see below).

The 4.4-m core from the Tahoe basin came from Osgood Swamp, a moraine-dammed depression. Logistics for coring an open water body were not clear as we began the work; we did not know whether we could wade or would need a boat or raft. When we reached the swamp, we found that it had been drained three weeks earlier by a local developer who hoped to control the mosquito population. Coring in the Sierra Nevada was off to an auspicious start. The Osgood Swamp record (Adam, 1967) was the first pollen record from the Sierra Nevada

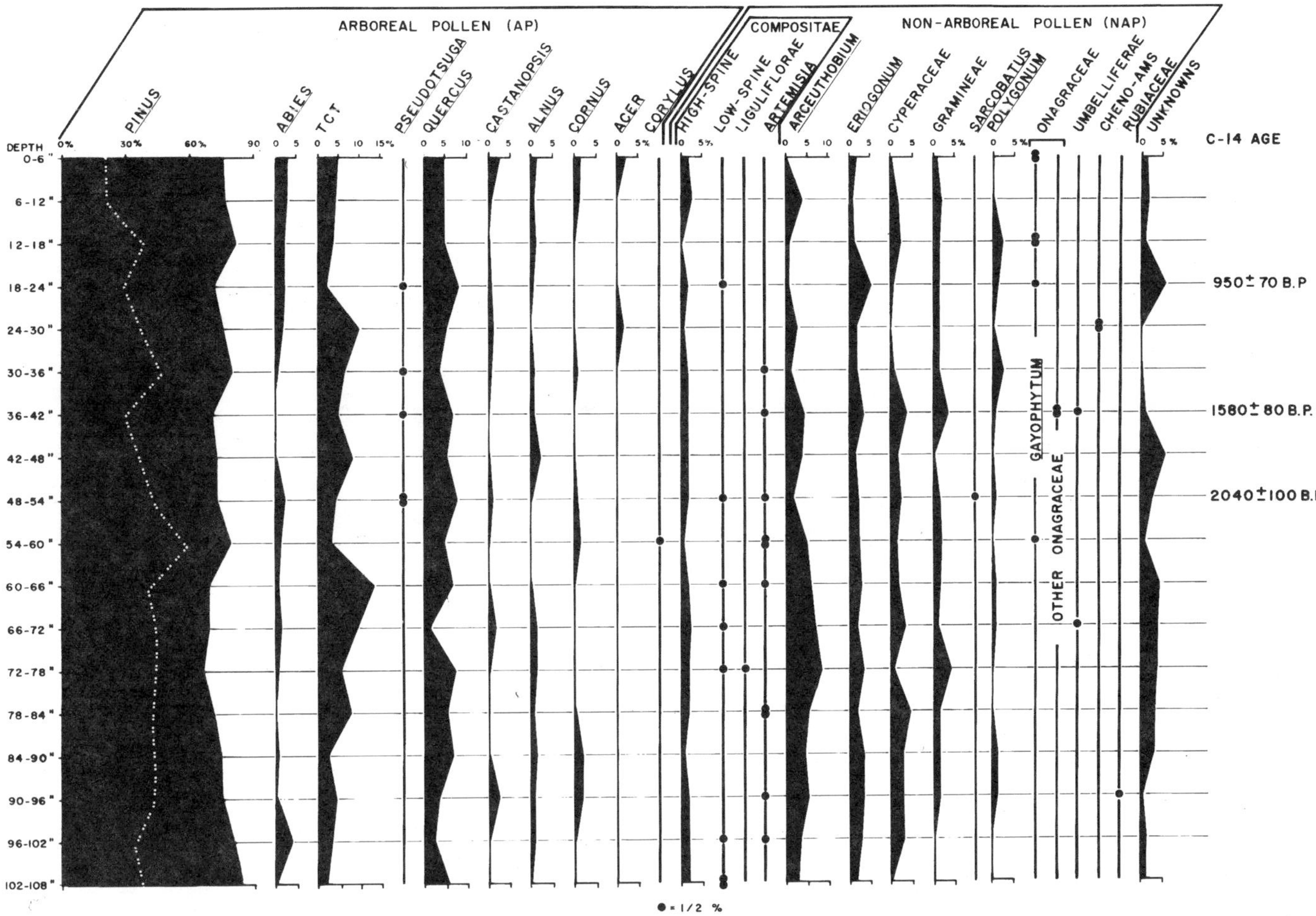

FIG. 1. Pollen diagram from archeological site MRP-105, Crane Flat, Yosemite National Park (from Adam, 1964).

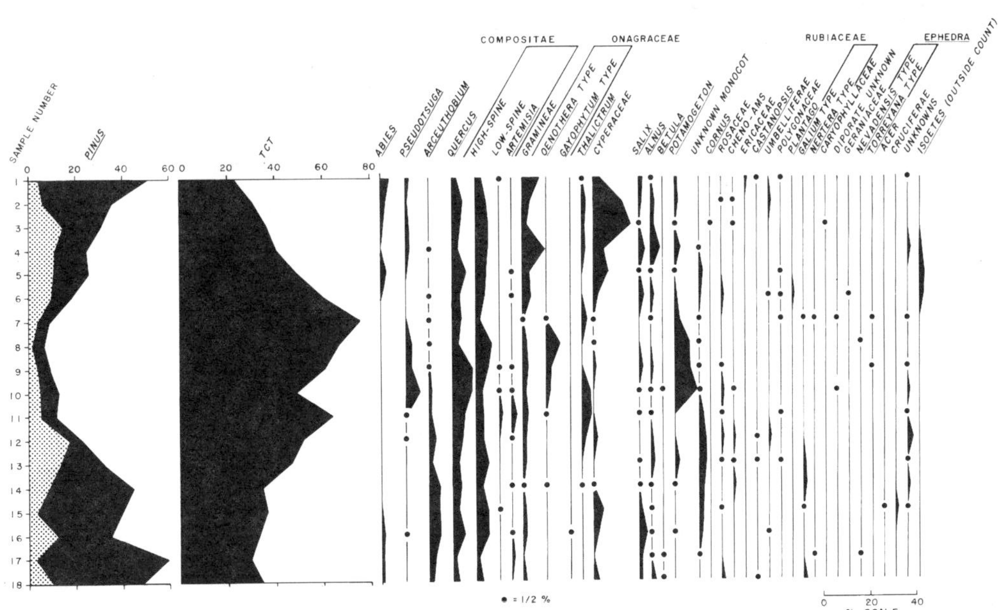

FIG. 2. Pollen diagram from alluvial section at Hodgdon Ranch, Yosemite National Park (from Adam, 1964). TCT pollen includes the families Taxodiaceae, Cupressaceae, and Taxaceae, and probably consists mostly of *Calocedrus* at this site.

to extend from the present back into the late Pleistocene.

EXTENSION INTO THE MOJAVE DESERT

The next step in the development of palynology in California was also done under the wing of Paul Martin, in the form of the Great April Fools' Day Expedition of 1964, led by P. J. Mehringer and named for the date of our departure from Tucson. Conventional wisdom at the time held that Holocene pollen records from the Mojave Desert would never be found because of lack of suitable deposits. Mehringer knew better, and organized the expedition to prove it by coring the deposits of playa-edge springs. The crew included Mehringer, Adam, the late Emma Lou Davis, Jim West, and Charlie Schweger. We recovered pollen-bearing middle to late Holocene cores from Warm Sulphur Springs in Panamint Valley, Saratoga Springs in southern Death Valley, and Zzyzx Hot Springs at Silver Lake. Mehringer was thus able to demonstrate the existence of Holocene pollen records from one of the most arid regions of North America.

Another early contribution to the pollen record at the Sierra/Mojave Desert boundary was the study of the pollen record from Black Lake, and again the work reflects the influence of Paul Martin. As a student at Arizona State University, George Batchelder visited Martin's Tumamoc Hill Laboratory in Tucson to learn the techniques of palynology. A field trip to the Sierra Nevada during his stay introduced him to coring, and at that time he selected Black Lake as the site for his dissertation (Batchelder, 1970).

A RESIDENT PALYNOLOGIST

California finally acquired a resident Quaternary palynologist when I joined the U.S. Geological Survey in Menlo Park in 1971. Previous workers came, they studied, but they did not stay. The initial rationale for a USGS palynologist was to study the pollen record along the San Andreas Fault as a way of documenting earthquake recurrence intervals. In retrospect, this was a very premature effort. Because the sides of the fault are moving past each other at a rate of several centimeters per year, the closed depressions along the fault are unlikely to be very old. Also, depressions with major strike-slip faults running through them are not the best places to establish reference sections. Once this became clear, a search for reference sections away from the San Andreas began.

It was also in the early 1970s that other palynologists began to pay attention to California. George Batchelder taught palynology and biology at San Francisco State University for several years, and his students have produced interesting studies from the Siskiyou Mountains (Miller *et al.*, 1976) and the Sierra Nevada (Dorland, 1980; Batch, 1977). Roger Byrne has been teaching palynology in the Geography Department at the University of California at Berkeley since 1973. His students have conducted a number of late Pleistocene and Holocene pollen studies in central California (*e.g.*, Peterson, 1980; Sullivan, 1982).

TOWARD LONGER RECORDS

A major step forward in California palynology came in 1973, when the developing Earthquake Program of the U.S. Geological Survey undertook coring of Clear Lake in the northern Coast Ranges, hoping to recover deformed sedimentary structures as a record of past earthquake activity. Several cores were taken under the leadership of John Sims. The record of deformed structures was disappointing, but the pollen record of the longest (115-m) core provided a complete record of the entire last glacial cycle (Adam, 1988). The record is unusual in that a single pollen curve

(*Quercus*, oak) shows most of the significant variability in the record (Fig. 3).

Correlations of the Clear Lake pollen record with other long records demonstrate that pollen records from California are relevant to the wider problem of global climatic changes. The Clear Lake oak pollen curve closely tracks changes in deep-sea oxygen isotope ratios and in long pollen records from Europe for the interval from the Eemian to the present. These correspondences indicate that the Clear Lake record may be interpreted as a local representation of at least a northern hemispheric climatic signal. The presence in both California and Europe of climatic fluctuations that match the wiggles in the deep-sea $\delta^{18}O$ curves, particularly for Stage 5 (Adam, 1988), implies that those changes must have also affected the North American mid-continent. It appears that the Sangamon interglacial of the mid-continent (which is defined on a weathering interval, not on a depositional sequence) corresponds to all of Stage 5 (Adam, 1988; Eyles and Williams, 1992). At Clear Lake, however, the pollen

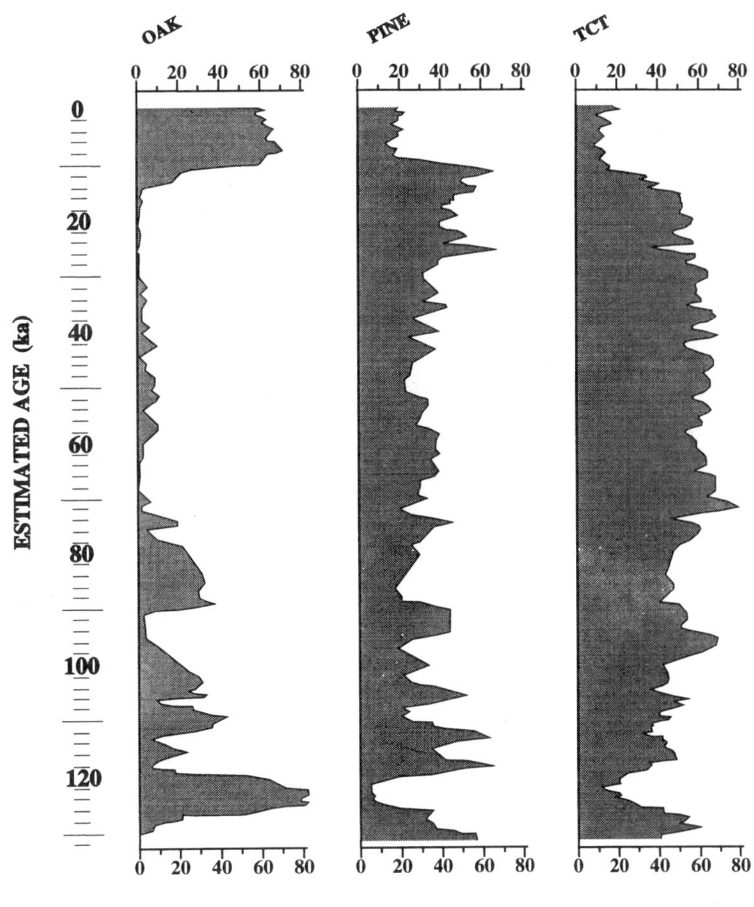

FIG. 3. Pollen diagram showing percentages of oak, pine, and TCT pollen *vs.* age for Clear Lake, core 4 (modified from Adam, 1988). TCT pollen includes the families Taxodiaceae, Cupressaceae, and Taxaceae, and probably consists mostly of *Calocedrus* and *Juniperus*.

record demonstrates that the local Stage 5 equivalent included an interval warmer than the present (the Konocti thermomer, Stage 5e equivalent), intervals a few °C cooler than present (the Boomli thermomers, Stages 5a and 5c equivalents), and intervals nearly as cool as the last full glacial interval (the Tsiwi cryomers, Stage 5b and 5d equivalents) (temperature estimates from Adam and West, 1983; thermomers and cryomers from Adam, 1988). Indeed, it seems likely to me that the Mono Basin and Tahoe glaciations of the Sierra Nevada correspond to the Tsiwi 1 and Tsiwi 2 cryomers; if so, then they are of Sangamon age (Adam, 1988). I conclude that we are dealing with a pathological nomenclature, and that the common practice of naming as "Sangamonian" or "Sangamon" any deposits that appear to represent the last interglacial is not helpful unless a direct correlation with the Sangamon soil is intended. Thus, the implications of the Clear Lake pollen record extend well beyond California.

In the present context it is appropriate to describe the correlation of the Clear Lake record with the pollen record recovered from Willcox Playa by Martin (1963). That record encountered difficulties typical when sampling the deposits of intermittent lakes in arid regions. These include wide variations in pollen frequencies over short intervals and periods of pollen destruction (by oxidation) or removal (by deflation) during times when the lake was dry. The project was among the first of its kind, and a bold step toward a Pleistocene record from southeastern Arizona's desert grasslands. Martin's initial interpretation of his record was generally correct, particularly considering that the numerous climatic wiggles within the "Sangamon" (I refer here to *all* of Stage 5) were basically unrecognized in North America at that time.

The Clear Lake pollen record, because of its detail and continuity, affords a more detailed interpretation of the Willcox Playa pollen record than was possible in 1963. The key to correlating the Willcox Playa record with the Clear Lake record lies in the oscillations between pollen-preserving and pollen-barren strata in the Willcox core. Martin interpreted the barren zones as representing pollen destruction by oxidation during periods of interglacial climate when the lake was dry (Martin, 1963). The proposed correlation between the Willcox record and the Clear Lake oak pollen curve (Fig. 4) shows that the pattern of oxidized *vs.* reduced zones in the Willcox core makes sense in terms of the oscillations observed in the Clear Lake equivalents to the deep-sea Stage 5 oscillations.

DEVELOPMENTS SINCE 1980

The period since 1980 has seen both an increase in the number of researchers and an increase in the time depth of studies. The Tucson pollen lab showed little interest in California pollen research between my departure in 1971 and the arrival of Owen Davis, who assumed command of the pollen research program in 1982. Since his arrival in Tucson, Davis and his students have been quite active in California, with studies in the Sierra Nevada, Mono Lake, Owens Lake, the Central Valley, and coastal marshes. One of his students, Scott Andersen, now teaching at Northern Arizona University in Flagstaff, leads a group actively engaged in pollen work in the Sierra Nevada (Anderson, 1986, 1990a, 1990b; Anderson and Carpenter, 1991; Anderson and Davis, 1988; Anderson *et al.*, 1985; Davis *et al.*, 1985; Smith and Anderson, 1992).

A REALLY LONG RECORD

The Willcox Playa pollen record is one of the earliest examples of western U.S. long pollen records acquired using drill rigs. The Clear Lake pollen study introduced large-scale drilling to California palynology. The next major

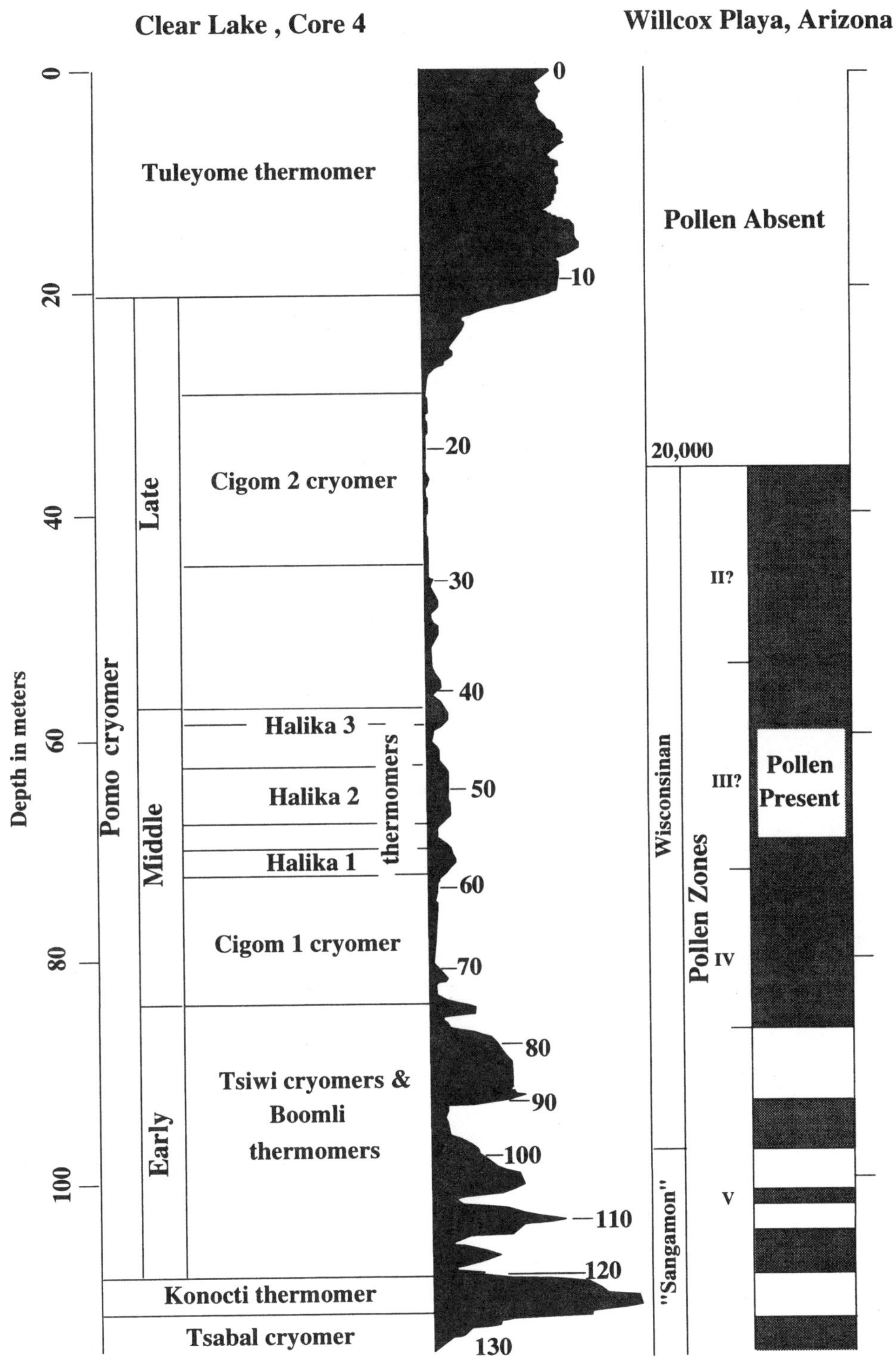

FIG. 4. Comparison of the Clear Lake, core 4, oak pollen curve (shaded) with the pollen presence *vs.* absence zones in the Willcox Playa pollen record of Martin (1963). Division of the Clear Lake record into informal climatic units is shown on the left, after Adam (1988). Estimated ages (ka) along the oak curve are shown by the numbers to the right of the curve. Age designations to the left of the Willcox Playa zones are those assigned by Martin (1963).

step forward was to drill deeper.

The opportunity for deeper drilling was provided by the Geothermal Program of the USGS, which was interested in the record of volcanic activity indicated by tephra deposits. The need for as complete a tephra record as possible led us to look to the region east of the Cascade Range, downwind from numerous volcanoes. The site selected was at the town of Tulelake in Siskiyou County, where a published water well log indicated a 550-m section of lake sediments. The site lies only about 20 km east of the shallow Klamath Lake sites analysed by Hansen (1942); this time around, however, we did more than just scratch the surface!

The Tulelake cores extended from the surface to a depth of 334 m (1100 feet), at which depth the drill rig was unable to lift the drill string safely. Drilling was done using a rotary wireline system, which permits core recovery without having to remove the entire drill string from the hole. The core has been studied for pollen, diatoms, paleomagnetism, geochemistry, and tephra (Adam *et al.*, 1989; Bradbury, 1991, 1992; Rieck *et al.*, 1992). Primary chronostratigraphic control for the core has been provided by a magnetic reversal chronology, coupled with tephra correlations to dated deposits elsewhere (Rieck *et al.*, 1992). The base of the core is about 3 million years old; with a larger drill rig, the record could be extended back even further.

In addition to providing one of the longest pollen records yet documented, the Tulelake core has established itself as one of the best sections available for tephrochronology in western North America. Over a dozen correlatable widespread silicic tephra layers have been identified in the core, and additional layers are present that have not yet been identified elsewhere. Because all of these tephra are present in a single stratigraphic sequence in known order, the core is particularly useful in dating tephra layers and in extrapolating those dates to other sites. The widespread distribution of identifiable tephra layers in upper Cenozoic deposits of western North America (Sarna-Wojcicki *et al.*, 1991) is a key element supporting the current surge of interest in coring and studying long records in this region.

FITTING THE PIECES TOGETHER

As California palynology embarks on its second half century, we have accumulated enough data to allow us to bring together initially unrelated studies to gain further insight into the evolution of the California pollen record. One such example is Adam and West (1983), which combined core-top pollen counts from short records in the California Coast Range with the long core records from Clear Lake to provide temperature and precipitation estimates through the past glacial cycle. A more recent example (Adam *et al.*, 1990) pointed out broad similarities between the climate dynamics operating at Tulelake during the Pliocene, at Clear Lake during the cooler parts of the last interglacial interval (*i.e.*, Stage 5), and at Hodgdon Ranch during the latter part of the Holocene.

The focus of the latter paper was on interpreting the part of the Tulelake Pliocene section older than 2 Ma. The two most important pollen types in this part of the section are pine and TCT (Fig. 5). A plot of the percentages of pine *vs.* TCT for the entire Tulelake section (Fig. 6A) showed a wide scatter of points with no particular pattern, but the cloud of points for the >2Ma samples (Fig. 6B) was much better defined, and offered the potential for interpretive insights. When the points for the >2Ma samples were connected by lines in stratigraphic order (Fig. 6C) the dynamics of the Tulelake Pliocene vegetation emerged: most lines have a negative slope, indicating that pine and TCT tended to replace each other in a reciprocal pattern, rather than increasing and decreasing together.

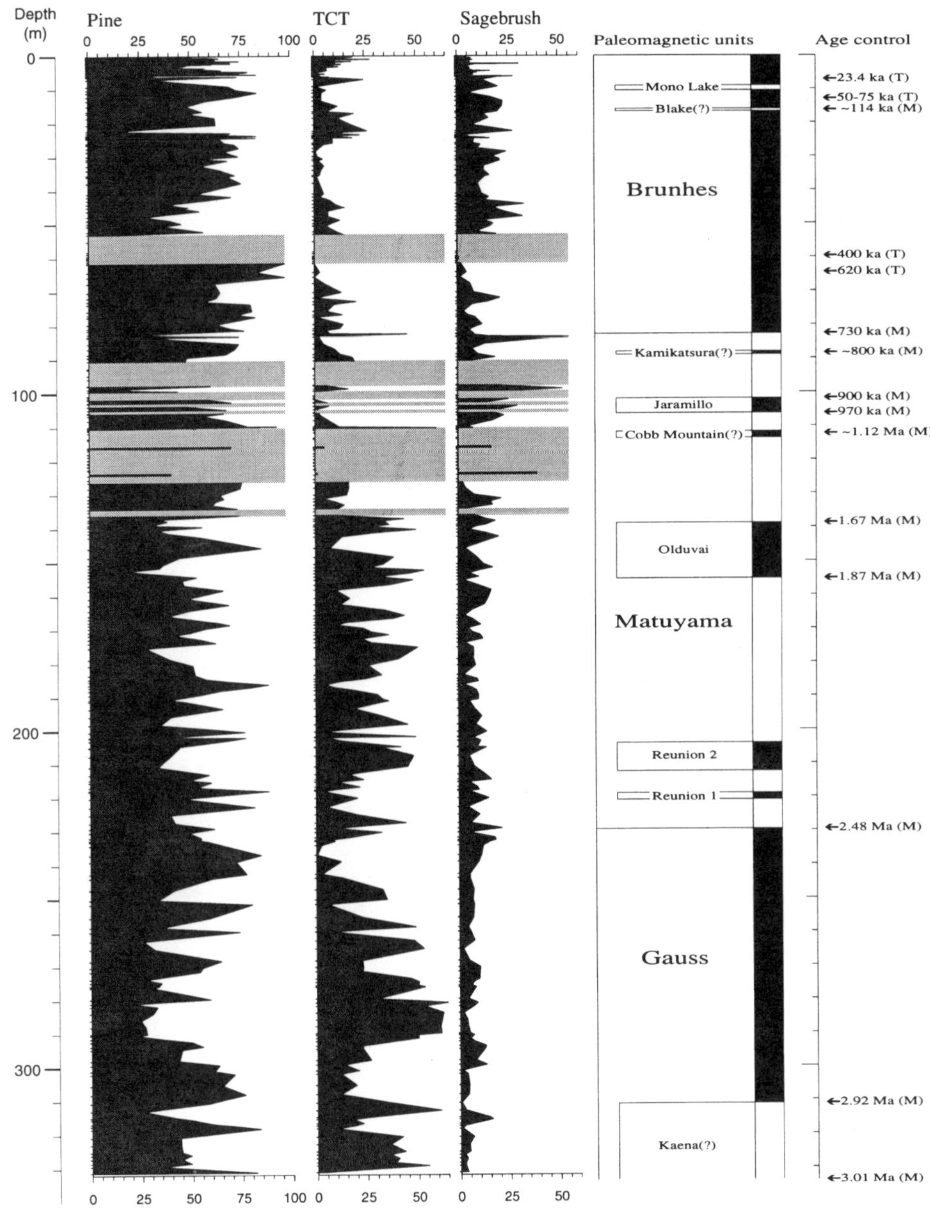

FIG 5. Summary percentage pollen diagram for the Tulelake section, slightly modified from Adam *et al.* (1990). TCT pollen includes the families Taxodiaceae, Cupressaceae, and Taxaceae, and probably consists mostly of *Calocedrus* and *Juniperus*. Shaded bars between 50 and 140 m represent intervals barren of pollen. Age calibration points are indicated by arrows along the right margin. Ages followed by (M) are based on published ages of paleomagnetic reversals; those followed by (T) are based on tephra of known age.

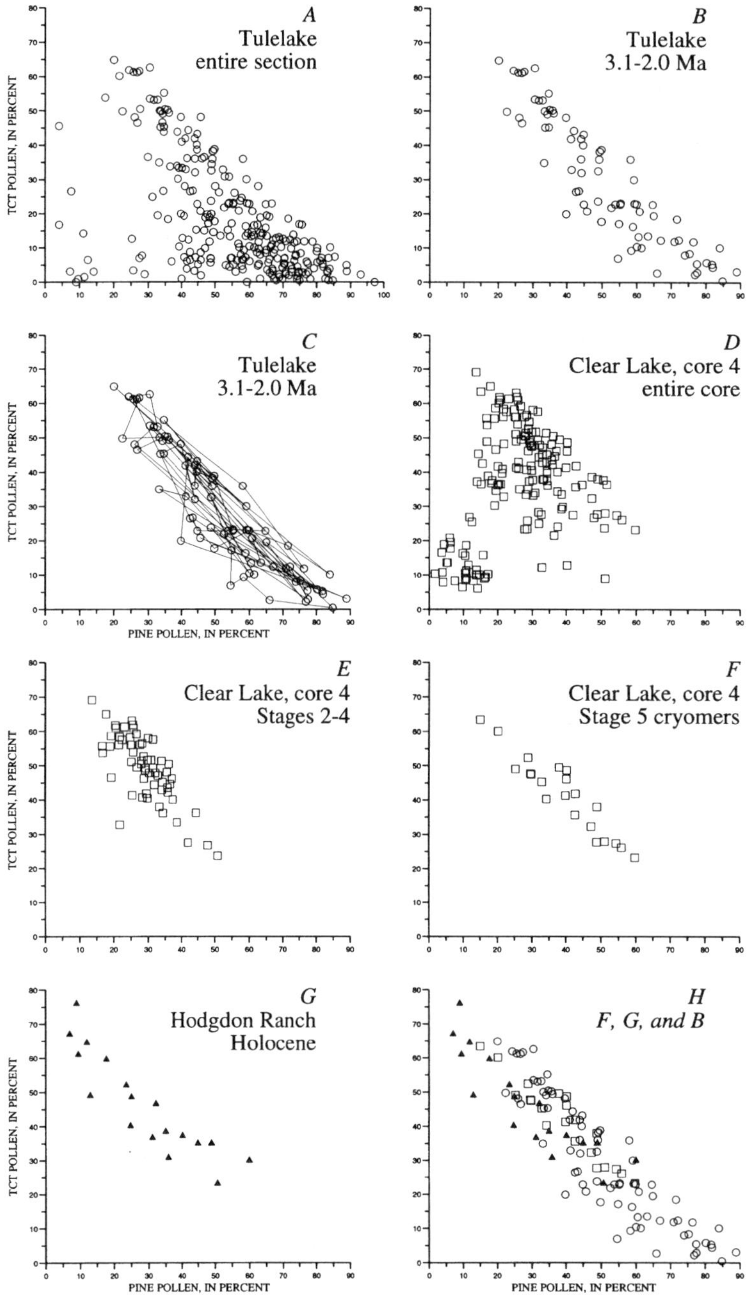

FIG. 6. Plots of pine *vs.* TCT percentages for selected data sets (see text for discussion). TCT pollen includes the families Taxodiaceae, Cupressaceae, and Taxaceae, and probably consists mostly of *Calocedrus* and *Juniperus*.

The next step was to see if such a pattern could be found elsewhere in the California pollen record. Plotting of the Clear Lake core 4 pine and TCT pollen data for the entire core (Fig. 6D) showed a diffuse scatter of points reminiscent of the plot for the entire Tulelake section (Fig. 6A), uninterpretable without more information. When the Clear Lake data points were plotted according to their stratigraphic units, however, the same general pattern observed for the Tulelake >2Ma samples emerged for both the deep-sea $\delta^{18}O$ Stage 2-4 equivalents (Fig. 6E, middle and late Pomo cryomer) and the deep-sea $\delta^{18}O$ substages 5b and 5d equivalents (Fig. 6F, Tsiwi cryomers), with the latter group providing the best match. Yet another plausible match (Fig. 6G) was found in the pollen record from Hodgdon Ranch, on the lower western slope of the Sierra Nevada. The superimposed patterns of Figures 6B, 6F, and 6G (H) clearly show the similarity between the records.

These comparisons involve only two pollen types, so the matches among the sites are true only in a general sense. The important point here is that enough pollen data have accumulated for California to allow us to begin to make such comparisons. We can now say with some confidence that the Pliocene climate dynamics at Tulelake were broadly similar to the dynamics at Clear Lake (400 m elevation in the Coast Ranges) during the cooler parts of deep-sea $\delta^{18}O$ Stage 5, and to the dynamics at Hodgdon Ranch (1300 m on the western slope of the Sierra) during the Holocene.

PRESENT WORK ON LONG RECORDS

As the likelihood of future global warming in response to human activities has acquired credibility in the past several years, the resources available for studies of climatic change have increased significantly. Several sites are currently being studied for their long pollen sequences, and the number of suitable sites for long-term records is large. Examples are described below.

The Upper Klamath Basin

The position of California, upwind of most of the U.S. and adjacent to the Pacific Ocean, has suggested to a number of workers that detailed correlations between the marine and continental paleoclimate records may be possible in this region, particularly considering the number of tephra layers already observed. A USGS workshop convened to evaluate the potential for such studies chose as its top continental priority a series of cores from the Klamath Falls area (Gardner *et al.*, 1991). Six sites were cored in 1991, four in Oregon and two in California. The sites vary in basin size, hydrography, vegetation, and degree to which they were glaciated. One California site, Butte Valley, produced a 102-m core that extends back into the Jaramillo Normal-Polarity Subchron of the Matuyama Reversed-Polarity Chron, or nearly 1 million years. The other, Grass Lake, appears to be 50,000 to 100,000 years old, and is being studied for pollen by Kathryn Hakala of the University of Pittsburgh.

The Mohawk "Lake Beds"

The northern Sierra Nevada, north of Lake Tahoe, contains numerous sediment-filled basins that offer good prospects for long climate records. One of these basins, near the hamlet of Mohawk on the upper Feather River, has been breached and deeply eroded, so that a 30-m section of sediment is exposed over a distance of about 350 m. The deposits include several tephra layers that indicate a time range for the deposits of about 160-750 ka. Normally-magnetized glaciolacustrine sediments are exposed at the base of the section, and glacial outwash gravels cap the section. Between those units lies a sequence of alternating peats, highly organic silts, and low-organic silts that can be grouped into 5 more or less cyclical packets. These beds have been sampled for pollen; all samples examined to date contain pollen, including more *Picea* (spruce) pollen than has been observed in Holocene samples from the Sierra Nevada.

The tephra units found in the section will permit precise correlations with the Tule Lake core and with the older cores from the upper Klamath Basin (Butte Valley, California, and Buck Lake, Oregon) at several horizons during the middle Pleistocene. Both the Mohawk section and the Buck Lake record are from heavily forested drainages, so the combined studies should offer a first look at the long-term dynamic behavior of the Cascade-Sierran forests spanning multiple glacial cycles.

Owens Lake

In 1992, the USGS recovered a new core from the deposits of Owens Lake under the leadership of G.I. Smith. The 300+ m core penetrates the Bishop Tuff near the base, which means that the core spans nearly all of the Brunhes Normal-Polarity Chron. The upper 65 m of the section, correlative with the last complete glacial cycle, is being studied for a Ph.D. dissertation by Wally Woolfenden under Owen Davis at Tucson, and the lower part of the section has been examined by USGS palynologists.

OBSERVATIONS ON THE FUTURE

Palynology was eliminated from the curriculum at San Francisco State several years ago. As a result, there is now only one university position in Quaternary palynology in California. Arizona, by contrast, has two programs that are active in training California palynologists. There are also active research programs involving California at both the University of Oregon under Cathy Whitlock and the Desert Research Institute under Peter Wigand. When one considers the size of the state of California and the diversity of its academic resources, the continuing lack of support for research into the vegetational and climatic history of the state is curious.

As to future research, there are many outstanding opportunities. A published pollen flora for all or significant parts of the California flora would contribute immeasurably to progress. Holocene sequences are becoming fairly common, particularly in the mountainous parts of the state, but late Pleistocene records are still scarce. Detailed and well-dated records from practically anywhere in the state would be valuable contributions.

Perhaps the most outstanding opportunity, and one that distinguishes California from much of the rest of the world, lies in the long records preserved in the western Great Basin, in the Central Valley, and in parts of the Sierra Nevada and the Coast Ranges. These records may span hundreds of thousands of years, and can provide climate records from western North America to match the much better-known deep sea $\delta^{18}O$ records. Such data can provide crucial ground-truthing for the output of the Global Climatic Models now being used to simulate both steady-state climates and climate dynamics. The relevance of the long California pollen record is thus easy to demonstrate. The students of Paul Martin have contributed much to its initial exploration, and his academic grandchildren and later descendants are hard at work. We hope to do him justice.

REFERENCES

Adam, D.P. (1964). Exploratory palynology in the central Sierra Nevada, California. Interim Research Report No. 4, Geochronology Department, University of Arizona, Tucson.

Adam, D.P. (1967). Late Pleistocene and Recent palynology in the central Sierra Nevada, California. *In* "Quaternary Paleoecology" (E. J. Cushing and H. E. Wright, Jr., Eds.), pp. 275-301. Yale University Press, New Haven.

Adam, D.P. (1985). Quaternary pollen records from California. *In* "Pollen Records of Late-Quaternary North American Sediments" (V. M. Bryant and R. G. Holloway, Eds.), pp. 125-140. American Association of Stratigraphic Palynologists Foundation, Dallas.

Adam, D.P. (1988). Palynology of two upper Quaternary cores from Clear Lake, Lake County, California. *U.S. Geological Survey Professional Paper* **1363**, 1-86.

Adam, D.P., Bradbury, J.P., Rieck, H.J., and Sarna-Wojcicki, A.M. (1990). Environmental changes in the Tule Lake basin, Siskiyou County, California, from 3 to 2 million years before present. *U.S. Geological Survey Bulletin* **1933**, 1-13.

Adam, D.P., and West, G.J. (1983). Temperature and precipitation estimates through the last glacial cycle from Clear Lake, California, pollen data. *Science* **219**, 168-170.

Adam, D.P., Sarna-Wojcicki, A.M., Rieck, H.J., Bradbury, J.P., Dean, W.E., and Forester, R.M. (1989). Tulelake, California: the last 3 million years. *Palaeogeography, Palaeoclimatology, Palaeoecology* **72**, 89-103.

Anderson, R.S. (1986). Late-Quaternary paleoecology of the central Sierra Nevada, California. *Crossosoma* **12**, 6-7.

Anderson, R.S. (1990a). Holocene forest development and paleoclimates within the central Sierra Nevada, California. *Journal of Ecology* **78**, 470-489.

Anderson, R.S. (1990b). Modern pollen rain within and adjacent to two giant sequoia (*Sequoiadendron giganteum*) stands, Yosemite and Sequoia National Parks, California. *Canadian Journal of Forest Research* **20**, 1289-1305.

Anderson, R.S., and Carpenter, S.L. (1991). Vegetation change in Yosemite National Park, California, during the protohistoric period. *Madroño* **38**, 1-13.

Anderson, R.S., Davis, O.K., and Fall, P. (1985). Late-glacial and Holocene vegetation change in the Sierra Nevada of California, with particular reference to the Balsam Meadow site. *In* "Late Quaternary Vegetation and Climates of the American Southwest" (B.F. Jacobs, P. Fall, and O.K. Davis, Eds.), pp. 127-140. American Association of Stratigraphic Palynologists, Contributions Series 16.

Anderson, R.S., and Davis, O.K. (1988). Contemporary pollen rain across the central Sierra Nevada, California: relationship to modern vegetation types. *Arctic and Alpine Research* **20**, 448-460.

Axelrod, D.I., and Ting, W.S. (1961). Early Pleistocene floras from the Chagoopa surface, southern Sierra Nevada. *University of California Publications in Geological Sciences* **39**, 119-194.

Batch, J.R. (1977). "A Post-Glacial Pollen Record from a Subalpine Valley in Yosemite National Park." Unpublished M.A. thesis, San Francisco State University.

Batchelder, G.L. (1970). "Post-Glacial Ecology at Black Lake, Mono County, California." Unpublished Ph.D. dissertation. Arizona State University, Tempe, Arizona.

Bradbury, J.P. (1991). The late Cenozoic diatom stratigraphy and paleolimnology of Tule Lake, Siskiyou Co. California. *Journal of Paleolimnology* **6**, 205-255.

Bradbury, J.P. (1992). Late Cenozoic lacustrine and climatic environments at Tule Lake, northern Great Basin, USA. *Climate Dynamics* **6**, 275-285.

Cressman, L.S. (1942). Archeological researches in the northern Great Basin. *Carnegie Institution of Washington Publication* ***538***.

Davis, O.K., Anderson, R.S., Fall, P.L., O'Rourke, M.K., and Thompson, R.S. (1985). Palynological evidence for early Holocene aridity in the southern Sierra Nevada, California. *Quaternary Research* **24**, 322-332.

Dorland, D. (1980). "Two Post-Glacial Pollen Records from Meyers Grade Marsh and Grass Lake, El Dorado County, California." Unpublished M.A. thesis, San Francisco State University.

Eyles, N., and Williams, N.E. (1992). The sedimentary and biological record of the last interglacial-glacial transition at Toronto, Canada. *In* "The Last Interglacial-Glacial Transition in North America" (P.U. Clark and P.D. Lea, Eds.), pp. 119-137. *Geological Society of America Special Paper* **270**.

Frink, J.W., and Kues, H.A. (1954). Corcoran Clay — A Pleistocene lacustrine deposit in San Joaquin Valley, California. *Bulletin of the American Association of Petroleum Geologists* **38**, 2357-2371.

Gardner, J.V., Sarna-Wojcicki, A.M., Adam, D.P., Dean, W.E., Bradbury, J.P., and Rieck, H.J. (1991). Report of a workshop on the correlation of marine and terrestrial records of climate changes in the western U.S. *U.S. Geological Survey Open-File Report* **91-140**, 1-48.

Hansen, H.P. (1942). A pollen study of peat profiles from Lower Klamath Lake of Oregon and California. *In* Cressman, L.S., Archeological researches in the northern Great Basin. *Carnegie Institution of Washington Publication* **538**, 103-114.

Heusser, C.J. (1960). Late-Pleistocene environments of North Pacific North America. *American Geographical Society Special Publication* **35**.

Martin, P.S. (1963). Geochronology of pluvial Lake Cochise, southern Arizona. II. Pollen analysis of a

42-meter core. *Ecology* **44**, 436-444.

Miller, J.S., Batchelder, G.L., and Lee, G.K. (1976). A postglacial pollen record from Campbell Lake, Siskiyou County, California. *American Quaternary Association, Abstracts of the Fourth Biennial Meeting, October 9 and 10, 1976*, 151-152.

Osvald, H. (1936). Stratigraphy and pollen flora of some bogs of the North Pacific coast of North America. *Berichtungen des Schweizerisches Botanisches Gesellschaft* **46**.

Peterson, S. (1980). "Late Holocene Sedimentation at Zaca Lake, Santa Barbara County, California." Undergraduate thesis on file at Geography Department, University of California (Berkeley).

Rieck, H.J., Sarna-Wojcicki, A.M., Meyer, C.E., and Adam, D.P. (1992). Magnetostratigraphy and tephrochronology of a 3-m.y. late Pliocene to Holocene record in lake sediments at Tulelake, Siskiyou and Modoc Counties, northern California. *Geological Society of America Bulletin* **104**, 409-428.

Roosma, A. (1958). A climatic record from Searles Lake, California. *Science* **128**, 716.

von Post, L. (1916). Om Skogstradpollen i sydsvenska torfmosslagerfoljder. *Geologiska Foreningens i Stockholm Forhandlingar* **38**, 384-394.

Sarna-Wojcicki, A.M., Lajoie, K.R., Meyer, C.E., Adam, D.P., and Rieck, H.J. (1991). Tephrochronologic correlation of upper Neogene sediments along the Pacific margin, conterminous United States. *In* "Quaternary Nonglacial Geology: Conterminous U.S." (R.B. Morrison, Ed.), pp. 117-140. The Geology of North America, vol. K-2.

Smith, S.J., and Anderson, R.S. (1992). Late Wisconsin paleoecologic record from Swamp Lake, Yosemite National Park, California. *Quaternary Research* **38**, 91-102.

Sullivan, D.G. (1982). "Prehistoric Flooding in the Sacramento Valley: Stratigraphic Evidence from Little Packer Lake, Glenn County, California." Unpublished M.A. thesis, University of California (Berkeley).

Chapter 10

Equable Climates, Mixed Assemblages, and the Regression Fallacy

KENNETH L. COLE

National Biological Survey CPSU, Department of Forest Resources
University of Minnesota, St. Paul, Minnesota 55108

INTRODUCTION

Mixed Assemblages

One of the longest running philosophical debates in paleoecological interpretation concerns the importance of mixed, or disharmonious, assemblages which represent past communities with no modern analog. These mixed assemblages challenge our world-view because they combine species which do not co-occur today in any known habitat, thus preventing easy classification.

Mixed assemblages are usually explained by invoking past climates more "equable" than that of today. An equable climate is typically envisioned as one in which the limiting extremes for individual species are reduced, usually in the form of warmer winters and cooler or wetter summers. Following its adoption to account for fossil records of giant tortoise in the American midcontinent (Hibbard, 1960), this mechanism has been used to explain disharmonious assemblages of plants (Van Devender and Burgess, 1985; Van Devender, 1990a and 1990b; Foster *et al.*, 1990), mammals (Graham and Mead, 1987; Graham and Semken, 1987), and herpetofauna (Holman, 1976). Because my experience is mostly botanical, I shall restrict this paper mainly to a discussion of fossil plants during the last 25,000 years, although similar logic might be applied to other organisms and time scales.

Individual scientists apparently have different tolerances for the amount of mixing that they will allow in their reconstructed associations. If the amount of disharmony in the fossil assemblages exceeds their conception of the inherent randomness of nature, then a past no-analog climate is used to explain the no-analog association. As a result, paleoclimatic interpretations can be influenced by an individual scientist's conception of order and randomness in nature.

Paleoclimatic reconstructions are also affected by the characteristics that a scientist believes to be most important in defining the association. Some would describe a past asso-

Late Quaternary Environments and Deep History: A Tribute to Paul S. Martin, edited by David W. Steadman and Jim I. Mead. The Mammoth Site of Hot Springs, South Dakota, Inc. Scientific Papers, Volume 3. Hot Springs, South Dakota, 1995.

ciation by considering the whole collection of species and finding the nearest approximate modern analog, even if no exact analog exists. Others derive their paleoclimatic estimates from a few species thought to be better indicators of past climates ("indicator" or "sensitive" species). The fundamental difference in interpretation can be thought of as defining an association by its "means" or by its "extremes". The analog approach is difficult when no close modern analog can be found. The indicator species approach is unreliable unless the physiological tolerances of these species are well established (rarely the case), and only climatic factors could have affected their distributions (almost never the case). The potential pitfalls of total reliance on this indicator species approach will be discussed in this paper.

THE VIEW OF THE PAST

To better understand the extent to which paleoenvironmental reconstructions are determined by our modern world-view, consider the world from the viewpoint of an imaginary Paleoindian botanist who arrived in North America around 18,000 years ago, during the late Wisconsinan, and classified the plant associations.

Of course, extra-continental travel is only one of the skills of our hypothetical Paleoindian botanist. He also has a touch of clairvoyance, and can see glimpses of future plant associations from our century. His glimpses of our time are similar to what we can see in fossil assemblages from his time.

Next the Paleoindian botanist attempts to describe our modern conditions based upon his understanding of full glacial plant associations and his glimpses of modern associations of plants. How would he describe our time? I suspect that many of our modern associations were rare or did not exist in his classification scheme, thus making them "anomalous" in his eyes. He would be forced to conclude that our time must be typified by comparatively equable climates in order to permit such strange mixtures of species.

For example, ponderosa pine (*Pinus ponderosa*) woodland is a major vegetation type today, extending throughout the western United States from the Sierra Madre Occidental of northern México to southern British Columbia in Canada (Peet, 1988; Little, 1971). In contrast, during the late Wisconsinan it was far less common (Anderson, 1989). Despite the analysis of over one thousand packrat middens from throughout the western United States (Webb, 1985), species-specific identifications of pine needles from packrat middens record only three fossil localities for ponderosa pine: Kings Canyon of the western Sierra Nevada (Cole, 1983), the Santa Catalina Mountains of southern Arizona (Van Devender, 1990a), and the San Andreas Mountains of southern New Mexico (Van Devender, 1990b). Although there are almost certainly other localities where it will be discovered, it appears that ponderosa pine is far more widely distributed today than it was during the late Wisconsinan.

Exercising his clairvoyant talents and comparing his familiar late Wisconsinan communities with the strange future assemblages with abundant ponderosa pine, what would our Paleoindian botanist make of it? It would be confusing for him to see that ponderosa pine and white fir (*Abies concolor*) now have similar distributions (Little, 1971). In many late Wisconsinan records white fir is abundant without a trace of ponderosa pine (Cole, 1990; Spaulding, 1990). Looking at specific future assemblages, he would have to conclude that our modern climate is equable if he were to see the Black Hills of South Dakota and find ponderosa pine growing with eastern species such as horizontal juniper (*Juniperus horizontalis*), eastern cottonwood (*Populus deltoides*), and peachleaf willow (*Salix amygdaloides*) (Little, 1971).

In the Eastern United States there are other

modern associations with no Pleistocene analog, as many of the modern associations in the Eastern Deciduous Forest have developed during the middle to late Holocene. For example, the hemlock-white pine-northern hardwoods forest of the upper Great Lakes region (Braun, 1950) has no known late Wisconsinan occurrence despite the fact that its characteristic species produce abundant pollen identifiable to genus. There are only isolated records of fossil pollen of basswood (*Tilia*), hemlock (*Tsuga*), beech (*Fagus*), and birch (*Betula*) in the late Wisconsinan, and the other major species of this northern hardwoods forest, white pine (*Pinus strobus*) and maple (*Acer*), were less abundant than today (Delcourt and Delcourt, 1987). If our Paleoindian botanist were to see the Great Lakes region today he most likely would conclude that our modern climates must be equable to permit the co-habitation of this broad area by these formerly more restricted species.

As is illustrated in Figure 1, the use of rigid classification schemes developed from one reference time period can bias the understanding of a different time period. Although Pleistocene climates may have indeed been more equable than those of today, it is unwise to reach this conclusion without an understanding of the statistical or other errors that could bias the analysis.

Although paleoecologists often conclude that past climates were equable, it is difficult to create equable climates in continental interiors using climate circulation models (Sloan and Barron, 1990). One possible reason for this contradiction is regression error.

THE REGRESSION FALLACY

The conclusion that past mixed assemblages indicate equable climates is subject to a statistical bias analogous to what has been termed the "regression fallacy". This problem, first noted by Sir Francis Galton (1822-1911) (Freedman *et al.*, 1991), arises when one uses changes in extreme members of a group to infer a change in the variability of the whole population. The resulting effect is often termed "regression toward the mean." However, because we interpret past species assemblages based upon knowledge of the present, in this context it might better be termed "retrospective regression toward the mean" or, alternatively, "regression away from the mean".

To demonstrate how this error can arise, consider the flawed logic of the following study on the effect of repeated testing on the variability of SAT (Scholastic Aptitude Test) scores. This example is analogous to estimating the variability within plant communities by considering changes in only the most extreme members.

A group of students all take the SAT exam twice in order to study the effect of repeated testing on the variability within the group. To determine whether a change occurred between the two tests, students are classified into three groups based upon their performance on the second test: 1) "Low" (0 to 29th percentile), 2) "Middle" (30th to 69th percentile), and, 3) "High" (70th to 99th percentile). Thus, the <u>second</u> test supplies our classification scheme of student groups which is analogous to the <u>modern</u> species distributions supplying our classification schemes of plant associations.

It is likely that some students improved their scores between the first and second test because factors other than pure aptitude (e.g. mood, luck, practice) impart variability into each individual score. Therefore, it is likely that some of the members of our "High" group (as defined by the second test) would have only ranked in the "Middle" had it been defined by the first test. These students, in the 59th through 69th percentiles on the first test, would have improved their scores to exceed the 70th percentile on the second test. The progress of those students who dropped from above to below the 70th percentile is ignored, as we are only interested in the "High" students as defined by the second test. As a result, the "high" students as defined by the second exam must have, as a group, averaged lower on the first exam. Again, the "High" group includes many who have increased their scores between the exams, but excludes those whose scores decreased to below the 70th percentile.

Similarly, some of the members of our "Low" group (as defined by the second test) would have ranked in the "Middle" had it been defined by the first test. These students, in the 30th to 40th percentiles on the first test, would have lowered their scores to below the 30th percentile on the second test. The progress of those students who improved to above the 30th percentile is ignored as we are only following the "Low" students as defined by the second test. As a result, the "Low" students as defined by the second exam must have, as a group, averaged higher on the first exam.

If we interpret these results in a manner analogous to some paleoclimatologists, we infer that the disparity between students, and the variability in the population, must have increased between the tests. Because the "High" students increased their average from the first test, while the "Low" students decreased their average, then conditions would seem to have been more equable, and the variability in the population less, during the first test. Of course, the true variability in the population has remained the same. It is only by following select populations of extreme students that this error in interpretation can arise.

Avoiding regression error in interpreting paleoenvironments

Avoiding regression error is difficult when paleoecological information is not complete enough to infer species absence (the so-called "negative" data) as well as presence. Imagine an assemblage of organisms adapted to a particular range of environmental values. Within this assemblage there are rare species and cosmopolitan species. A rare species can be thought of as an "extreme" species. It is limited to a unique sub-environment and is considered to be a sensitive indicator species. Its presence in a fossil assemblage is considered meaningful. In contrast, the cosmopolitan species can be thought of as an "average" species. Its presence in an assemblage is not very meaningful to an ecologist attempting to define the environmental tolerances within the assemblage, but its absence could be very meaningful.

To reconstruct past climates from indicator species, the paleoclimatologist must not only assume that the species tolerances are known, but further, that average species have remained average and rare species have remained rare. If any variable other than climate has caused a species to become more rare or more common, then paleoclimatological reconstructions based only on species presence will be subject to an increasing regression error based upon changes due to non-climatic variables (Fig. 1).

An area with an equable climate should contain plant species typical of equable climates, rather than just mixed assemblages. Any proposed past equable climate becomes suspect if there is a conspicuous absence of the species adapted to modern equable climates.

For example, southern Arizona has been described as having had equable Pleistocene climates on the basis of mixed assemblages (Van Devender, 1990a). This area is adjacent to coastal southern California, which has a modern equable climate. The proposed equable Pleistocene climate would be suspect if none of the modern cosmopolitan species that dominate the modern equable zones were present. A few of the desert-adapted chaparral species, such as Whipple yucca (*Yucca whipplei*) and California buckwheat (*Eriogonum faciculatum*), are present in the southern Arizona Pleistocene records (Van Devender,

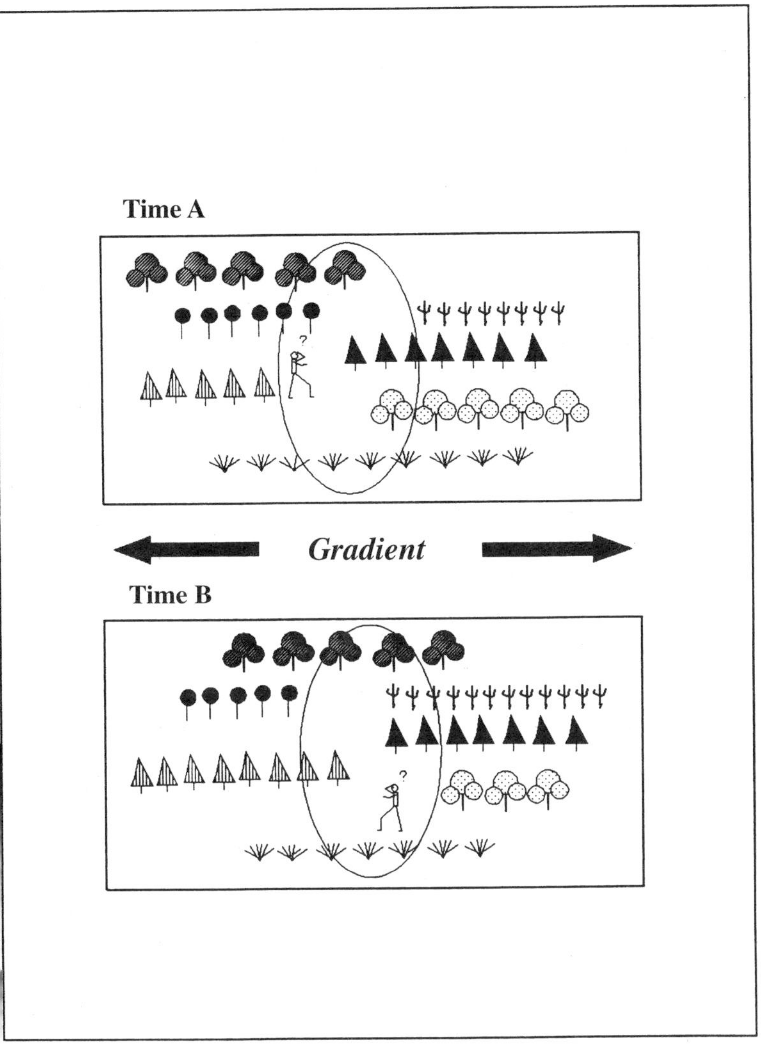

FIG. 1. Seven plant species are distributed along a gradient in potential climate space at two different times. Because physiographic gradients do not allow for the complete expression of the potential climate gradient, only that portion within the ovals can be observed during the two time periods, A and B. Although the plant species have undergone random shifts, contractions, and expansions between time A and B, climate within the two ovals is identical. Two observers, one within each oval, define the "modern" plant association of their oval based upon what they see. Because both observers define their vegetation types based upon their own "modern" observations, each interprets the vegetation of the other time as being "mixed" because it contains two species not typical of their own community classification.

1990a; Cole, 1986), supporting the proposed equable climate. Also, many of the higher elevation records contain abundant fossils of shrub oak which are similar to the oak species of the California chaparral. However, there are few records of the major sclerophylous, chaparral taxa that dominate the assemblages in the modern equable climatic areas. Except for the desert-adapted *Cercocarpus intricatus*, fossils of manzanita (*Arctostaphylos* spp.), *Adenostoma* spp., mountain mahogany (*Cercocarpus* spp.), and California lilac (*Ceanothus* spp.) are either rare or unknown. If the late Wisconsinan climate of southern Arizona was equable, it could not have been as equable as the modern climate west of the California coast ranges. The absence of the modern chaparral dominants sets a limit for the proposed Pleistocene equability.

Non-climatic factors in plant distribution

Some of the non-climatic variables that can affect species distributions and possibly lead to an error in the climatic interpretation of a fossil are listed in Table 1. I will not impugn the reputation of paleoecologists (especially myself) by claiming that any of these factors, such as contamination, are frequent. The point is that if any of these factors occur, then paleoclimatic reconstructions will inevitably be biased toward past equable climates. Because only the presence of extreme species is considered, it is unlikely that paleoclimatic reconstructions would erroneously suggest more extreme paleoclimates. By focusing on those species which are now extreme (rare), as in focusing on the extreme students from the second test, we are ignoring the progress of those species which are now considered average (cosmopolitan), but in the past were more extreme.

TABLE 1. Non-climatic factors that could bias results toward a past equable climate.

I. Incomplete climatic fidelity in modern distributions.

A. The modern population may not be in equilibrium with climate as measured by historical climate records. While a species' climatic limits are defined by the climatic values observed during the last 50 to 100 years, the species may be in the process of migrating and competing for its final equilibrium position. For example, many populations of longer-lived trees were established prior to the end of the Little Ice Age, and their distributions may more accurately reflect the climatic conditions of that time.

B. Severe vegetation changes have been recorded in historic time, most of which are the result of anthropogenic disturbances. If these disturbances have affected species distributions, then our conception of climatic tolerances will be biased. In North America, for example, severe disruptions accompanied settlement by Anglo-Europeans. Early changes (prior to 1850 A.D.), such the as effects of the introduction of *Equus* to the Great Plains in the 1700's, often are not well understood.

C. Past chance dispersals probably influence species ranges. The introductions of extra-continental exotic weeds which subsequently flourish demonstrate that climate is not the only determinant of vegetation dispersal. Similar chance dispersals within a continent could be responsible for much of the geographic distribution of plants, just as rare, long-range dispersals are thought to influence island floras.

D. Faunal interaction can influence plant distributions. Animals can act as herbivores, dispersers, and pollinators. Massive faunal changes, such as the megafaunal extinctions occurring in North America at the end of the Pleistocene, must have influenced plant distributions.

II. Genetic change. The climatic requirements of a species may have shifted since the time of the fossil assemblage. Although this seems unlikely within the Holocene, the rates of evolution for the climatic tolerances of species are not yet established.

III. Contamination of fossil assemblage with materials of different ages. Assemblages may become contaminated at many points in the process of study, including prior to their collection. For example, redeposition is possible in both cave and lake deposits.

IV. Misidentification of a fossil. Presumably all extreme unexpected members of an assemblage are carefully checked, although it is possible for a past assemblage to contain a fossil from an unknown past species or population that is not easily distinguishable from the modern species.

V. Incomplete knowledge of modern associations. Only those who have never been surprised by discovering an unexpected association in the field can deny this possibility.

CONCLUSIONS

In paleoecology, the present will always be used as the key to the past, as our understanding of most organisms and processes must be from the present. However, using a modern classification scheme to interpret the past requires that we fully understand the inherent biases produced by this method. One of these biases is in estimating changes in the variability within assemblages by comparing the past distributions of a few species to their modern distributions. Such an analysis will always be vulnerable to the regression fallacy. Although a change in variability within associations may be real, this cannot be determined unless the history of all of the assemblage members is tracked to determine whether some modern cosmopolitan species were formerly more restricted.

A more direct approach to recognizing past equable climates is to count those species that now reside in adjacent areas with equable climates. If the modern species typical of equable climates are absent in the paleoecological record, then the proposed equable paleoclimate becomes tenuous.

ACKNOWLEDGMENTS

This paper began sometime around 1979 as a few hastily scribbled notes while I was lucky enough to have a place at the Desert Laboratory on Tumamoc Hill. It could not have been started without the atmosphere of free thinking and critical examination of basic concepts inspired by Paul S. Martin. Putting these abstract thoughts into words was aided by the editing of Robyn Flakne, and critical reviews from Geoffrey Spaulding, Paul S. Martin, Gary Oehlert, and Peter Van de Water.

REFERENCES

Anderson, R. S. (1989). Development of the southwestern ponderosa pine forests: What do we really know? *In* "Multiresource management of Ponderosa Pine forests, Rocky Mountain Forest and Range Experiment Station," General Technical Report RM-185, 15-22.

Braun, E. L. (1950). "Deciduous forest of Eastern North America." Macmillan Publishing Co., New York.

Cole, K. L. (1983). Late Pleistocene vegetation of Kings Canyon, Sierra Nevada, California. *Quaternary Research* **19**, 117-129.

Cole, K. L. (1986). The lower Colorado River valley: a Pleistocene desert. *Quaternary Research* **25**, 392-400.

Cole, K. L. (1990). Late Quaternary vegetation gradients through the Grand Canyon, *In* "Packrat Middens: the Last 40,000 Years of Biotic Change" (J.L. Betancourt, T.R. Van Devender, and P.S. Martin, Eds.), pp. 240-258. University of Arizona Press, Tucson, AZ.

Delcourt, P.A., and Delcourt, H.R. (1987). "Long-Term Forest Dynamics of the Temperate Zone." Springer-Verlag, New York.

Foster, D. R., Schoonmaker, P.K., and Pickett, S.T.A. (1990). Insights from paleoecology to community ecology. *TREE* **5**,119-122

Freedman, D., Pisani, R., Purves, R., and Adhikari, A. (1991). "Statistics." W.W. Norton and Co., New York, New York.

Graham, R. W., and Mead, J. I. (1987). Environmental fluctuations and evolution of mammalian faunas during the last deglaciation in North America. *In* "North American and Adjacent Oceans During the Last Deglaciation" (W.F. Ruddiman and H.E. Wright, Eds.), pp. 371-402. The Geology of North America, vol. K-3. Geological Society of America.

Graham, R. W., and Semken, H. A., Jr. (1987). Philosophy and procedures for paleoenvironmental studies of Quaternary Mammalian Faunas. *In* "Late Quaternary Mammalian Biogeography of the Great Plains and Prairies" (R.M. Graham, H.A. Semken, and M.A. Graham, Eds.). Illinois State Museum, Scientific Papers 22.

Hibbard, C. W. (1960). An interpretation of Pliocene and Pleistocene climates in North America. *Annual Report Michigan Academy of Science, Arts, and Letters* **62**, 5-50.

Holman, J. A. (1976). Paleoclimatic implications of "ecologically incompatible" herpetologic species. *Herpetologica* **32**, 290-295.

Little, E. L. (1971). Atlas of United States Trees, Vol. 1. U.S.D.A. Miscellaneous Publication 1146, Washington, D.C.

Peet, R. K. (1988). Forests of the Rocky Mountains. *In* "North American Terrestrial Vegetation" (Barbour, M.G. and Billings, W. D., Eds.). Cambridge University Press, New York.

Sloan, L. C., and Barron, E. J. (1990). "Equable" climates during earth history? *Geology* **18**, 489-492

Spaulding, W. G. (1990). Vegetational and climatic development of the Mojave Desert: the last glacial maximum to the present. *In* "Packrat Middens: the Last 40,000 Years of Biotic Change" (J.L. Betancourt, T.R. Van Devender, and P.S. Martin, Eds.), pp 166-199. University of Arizona Press, Tucson.

Van Devender, T. R. (1990a). Late Quaternary vegetation and climate of the Sonoran Desert, United States and Mexico. *In* "Packrat Middens: the last 40,000 Years of Biotic Change" (J.L. Betancourt, T.R. Van Devender, and P.S. Martin, Eds.), pp. 134-163. University of Arizona Press, Tucson.

Van Devender, T. R. (1990b). Late Quaternary vegetation and climate of the Chihuahuan Desert, United States and Mexico. *In* "Packrat Middens: the Last 40,000 Years of Biotic Change" (J.L. Betancourt, T.R. Van Devender, and P.S. Martin, Eds.), pp. 104-134. University of Arizona Press, Tucson.

Van Devender, T. R., and Burgess, T. L. (1985). Late Pleistocene woodlands in the Bolson de Mapimi—a refugium for the Chihuahuan Desert Biota. *Quaternary Research* **24**, 346-353.

Webb, R. H. (1985). Spatial and temporal distribution of radiocarbon ages of rodent middens from the southwestern United States. *Radiocarbon* **28**, 1-8.

Chapter 11

Environmental Change, Ecosystem Responses, and the Late Quaternary Development of the Mojave Desert

W. GEOFFREY SPAULDING

Dames & Moore, Inc., 4220 S. Maryland Parkway, Suite 108, Las Vegas, Nevada 89109

INTRODUCTION

Paleoecology is, like ecology itself, a study of interconnections and interactions. Well dated records of the responses of different systems to late Quaternary climatic change provide us with the opportunity to understand better the nature of these ecosystems, and the opportunity to begin teasing apart cause from effect, and effect from mere correlation. This, then, is the rationale for the rather pretentious title of this paper. This effort will also serve to highlight what we do not know- of chronologies, of the sensitivities of various systems, and of the causal chains that link those systems. Often the best that can be done is to point to apparent correlations, and speculate on possible causation.

In this paper I first discuss well-dated physical records that are, at most, only indirectly connected with the biotic "responses" evident in the paleoecological record. Second, I examine the evidence for ecological responses in the paleobotanical record available from ancient packrat middens and, to a lesser extent, from fossil pollen profiles. The first trophic level, composed of sessile autotrophs (green plants anchored in place), ought to offer the most unequivocal evidence of ecological responses to climatic change. Whether these are rapid responses and, therefore, demonstrate that vegetation was in "equilibrium" with climate, or whether there were biotically mediated lags in response times is a question that remains open (see for example, Cole, 1985; Huntley and Webb, 1989; Ritchie, 1986; Spaulding, 1990a). Finally, with these data in mind, I very briefly consider the chronology and nature of changes at the second and third trophic levels. Vertebrate extinctions at the end of the last glacial age need to be considered from a "bottom-up" point of view, where the cascade of effects forced by environmental change may be evaluated as an integrated system.

THE MOJAVE DESERT

The effects of past variations in effective moisture are evident in the raised shorelines

Late Quaternary Environments and Deep History: A Tribute to Paul S. Martin, edited by David W. Steadman and Jim I. Mead. The Mammoth Site of Hot Springs, South Dakota, Inc. Scientific Papers, Volume 3. Hot Springs, South Dakota, 1995.

and subsurface stratigraphy of dry lakes, in spring and marsh deposits that reflect higher groundwater levels and(or) increased discharge, and in relict soils. The evidence considered here comes from localities that lie between the southern Sierra Nevada and Transverse Ranges (the San Gabriel, San Bernardino, and San Jacinto Mountains) on the west and the longitude of the lower Colorado River on the east, and between latitudes 34°N and 37°30'N (Fig. 1). This area comprises the major portion of the Mojave Desert of Nevada and adjacent California, and is one of the most arid regions in North America. The reason for selecting such a discrete bioclimatic region is well-known to most paleoclimatologists. Things happen at different times in different areas (*e.g.*, Wright, 1984; Barnosky *et al.*, 1987), and by considering the evidence from a limited area, I reduce the probability of introducing geographic variations in the timing of climatic changes. At least theoretically, relatively modest changes in effective moisture should result in marked biotic responses in this area where effective moisture is so limited.

Climate of the Mojave Desert

The Mojave Desert is a warm temperate desert, typified by long, hot summers and brief but cool winters. The variance of average annual temperatures (T_a) from south to north through the region is strongly affected by the increasing elevation of the valley floors with increasing latitude (*e.g.*, Beatley, 1975). Although the Mojave Desert is not known as one that has a monsoonal precipitation regime, like the Sonoran and Chihuahuan Deserts, it is something of a mistake to think that the region is summer dry. Roughly 40% of the average annual rainfall (P_a) received in Las Vegas, Nevada, comes during the summer months. However, with such a low annual average ($P_a \approx 105$ mm) what summer rainfall that does fall in the valleys does little to mitigate the Mojave Desert's biotically critical summer drought. The relative amount of summer precipitation decreases from east to west across the Mojave Desert, with Barstow, California receiving only about 3% of its average annual increment of precipitation during the summer.

The major source of precipitable moisture aloft during both the summer and the winter is the eastern Pacific Ocean. Low pressure centered over the Mojave and Great Basin Deserts during the winter as a result of migratory low pressure systems, or during the summer as a thermal low, creates southwesterly winds aloft that bring maritime tropical air into the interior. This southwesterly flow of air skirts the major orographic barrier of the southern Sierra Nevada. Advected air during the summer months can describe a southerly flow from the Gulf of California.

Mojave Desert vegetation

The Mojave Desert is the smallest of the North American desert regions defined by Shreve (1942). Its general topography is similar to that of the rest of the Basin And Range Province: isolated mountain ranges surrounded by vast, arid valleys. Most Mojavean mountains are small and only four ranges have considerable topography above ca. 2200 m, the approximate lower limit of montane forest in this region. These are the Sheep and Spring Ranges of Nevada, and the Panamint Range and Inyo Mountains of California (Fig. 1). The highest peaks of these mountains reach over 3000 m, and support subalpine woodland dominated by Intermountain bristlecone pine (*Pinus longaeva*) and, on drier sites, black sagebrush scrub (*Artemisia nova*). True alpine vegetation does not exist on the mountains of the Mojave Desert.

It is of interest that the montane forest zone that is to be expected below subalpine woodland occurs on only two of these four large mountain ranges. The Panamint Range and

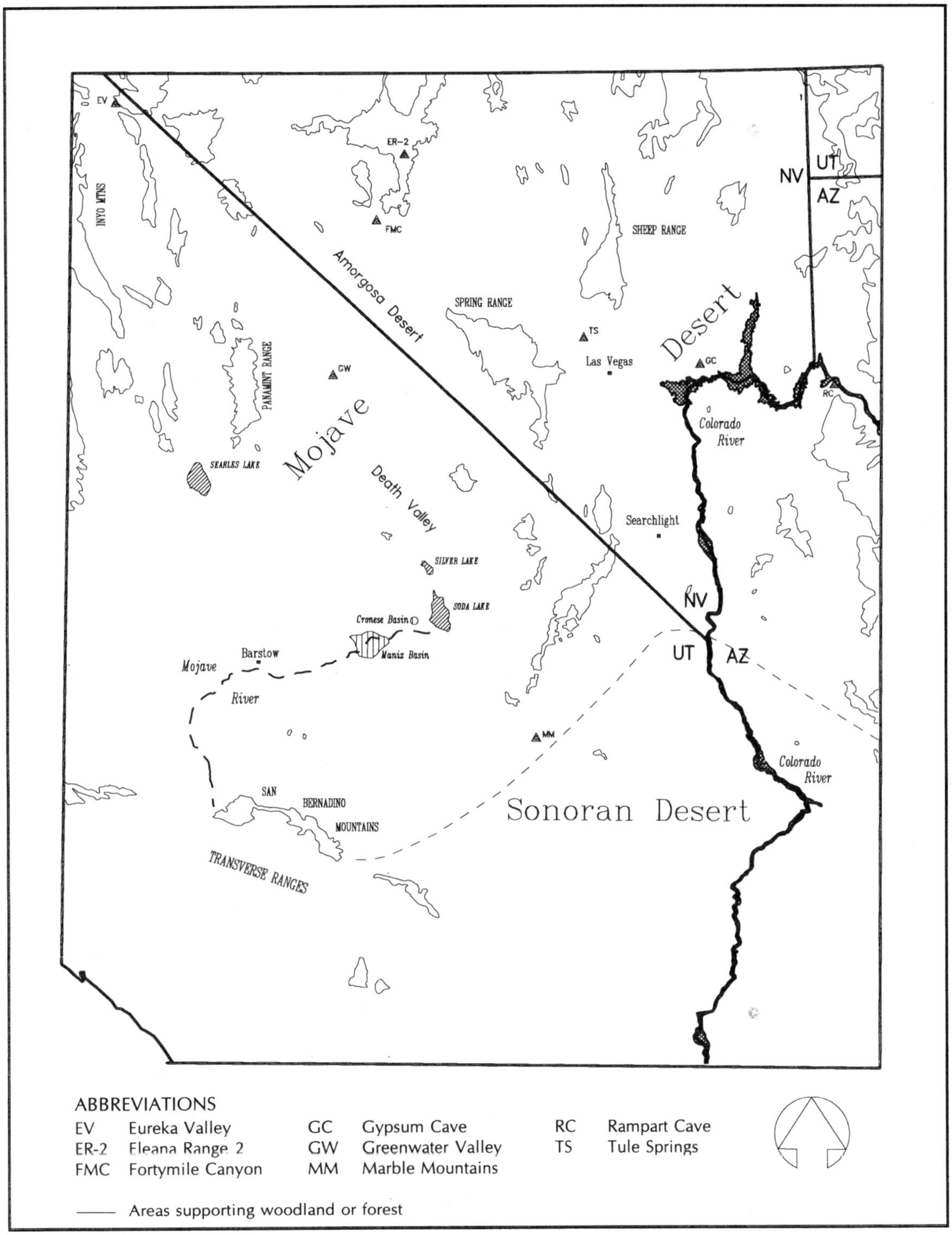

FIG. 1. The Mojave Desert with localities mentioned in the text.

Inyo Mountains do not support fir-pine forest (*Abies concolor-Pinus ponderosa*), while the Spring and Sheep Ranges do have substantial forest stands. Although it is possible that the absence of forest from the California ranges might be due to a lack of suitable highland corridors that would have allowed the postglacial immigration of forest trees to them (see Anderson, 1990, for a discussion of the postglacial migration of ponderosa pine), it is more likely that the rain shadow in the immediate lee of the massive Sierra Nevada results in a climate that is too dry to support these moisture-loving conifers. Elsewhere, in the Spring and Sheep Range of southern Nevada, fir-pine forest extends from about 2200 to 2750 m on mesic slopes. On more xeric aspects interior chaparral (Brown *et al.,* 1979) dominated by greasebush (*Forsellesia nevadensis*), little-leaf mountain mahogany (*Cercocarpus intricatus*), joint-fir (*Ephedra viridis*), and (in the Spring Range only) manzanita (*Arctostaphylos pungens*) extends over this elevational range, and down to lower slopes that are too xeric to support pinyon-juniper woodland.

Pinyon-juniper (*Pinus monophylla-Juniperus osteosperma*) woodland extends down to about 1830 m on the larger mountain ranges of the Mojave Desert, and also occurs on some of the smaller mountains. Oak (chiefly *Quercus turbinella* and *Q. gambelii*) is an uncommon woodland component, being restricted to the mountains near the southeastern periphery of the Mojave Desert. Other common components of Mojave Desert woodland include black sagebrush, cliff-rose (*Cowania mexicana*), joint-fir, and skunk-bush (*Rhus trilobata*). Near the lower ecotone with blackbrush desert scrub (*Coleogyne ramosissima*) such species as Joshua-tree (*Yucca brevifolia*), banana yucca (*Y. baccata*), and desert almond (*Prunus fasciculata*) are common components of woodland vegetation.

As the transition to blackbrush desert scrub is approached pinyon usually drops out before juniper. In the southern Mojave Desert, California juniper (*Juniperus californica*) can play an important role in plant communities of the upper bajada and lower mountain slopes. Two arborescent species of higher-elevation desert scrub that are often considered the hallmark of the Mojave Desert are Joshua-tree and Mojave yucca (*Yucca schidigera*). Less conspicuous desert perennials that are common in blackbrush scrub include ground-thorn (*Menodora spinescens*), desert-rue (*Thamnosma montana*), and ratany (*Krameria parvifolia*). Many of these species continue into lower-elevation mixed desert scrub as blackbrush and Joshua-tree decline in importance below about 1400 m elevation. On calcareous soils of the northern Mojave Desert such plants as spiny sagebrush (*Artemisia spinescens*), shadscale (*Atriplex confertifolia*), and Mojave pepperweed (*Lepidium fremontii*) play important roles in mixed desert scrub. To the south on igneous substrate, however, these species are often absent and, instead, shrubs such as the goldenweeds (*Haplopappus cuneatus*, *H. linearifolius*), Mojave sage (*Salvia mohavensis*), and the boxthorns (*Lycium cooperi*, *L. andersonii*) are common. Throughout the Mojave Desert creosote bush (*Larrea divaricata*) and Nevada ephedra (*Ephedra nevadensis*) are important elements in mixed desert scrub vegetation.

Below about 1000 m elevation creosote bush-white bursage (*Ambrosia dumosa*) desert scrub dominates. This is the same vegetation type that typifies low elevations of the Sonoran Desert, but many of the associated species are different and include Mojave caper (*Isomeris arborea*), California ephedra (*Ephedra californica*), California boxthorn (*Lycium californicum*) and, on fine-grained alkaline soils, desert holly (*Atriplex hymeneletra*). On rocky slopes within the creosote bush-white bursage zone such distinctive species as desert spruce (*Peucephyllum schottii*), rock nettle (*Eucnide urens*), and brittle-bush (*Encelia *farinosa*) can be found.

PHYSICAL RECORDS OF LATE QUATERNARY CLIMATIC CHANGE

Chrono-stratigraphic terms are used here to distinguish portions of late Quaternary time. The boundaries are arbitrary in the sense that a major climatic change can take place during one of these periods, rather than marking the boundary between two periods. I follow international convention in accepting a 10,000 yr B.P. date for the Pleistocene-Holocene boundary, synonymous with the late Wisconsinan-Holocene boundary (Olausson, 1982; Hopkins, 1975). The full-glacial, or Wisconsinan maximum, is dated at about 18,000 yr B.P. (CLIMAP, 1981), and the period between 23,000 and 10,000 yr B.P. encompasses the late Wisconsinan. The early Holocene is set at 10,000 to 7500 yr B.P., the middle Holocene from 7500 to 4000 yr B.P., and the late Holocene from 4000 yr B.P. to present.

At one point in time considerable energy was devoted debating the timing of the end of the last (Wisconsinan) glacial age, an issue that is now largely obviated by well-controlled deep-sea core and terrestrial records of deglaciation. They show that the wasting of the massive continental ice sheets that dominated the climate of the Northern Hemisphere began by ca. 14,000 yr B.P., and was completed by ca. 7500 yr B.P. (*e.g.*, Mix, 1987). Records from the Rocky Mountains and the western Cordillera suggest that the decay of the continental ice sheets lagged behind that of montane glaciers. Most montane areas were ice free by 12,000 yr B.P. (*e.g.* Andrews *et al.*, 1975; Porter *et al.*, 1983).

Lakes in the Mojave Sink

The headwaters of the Mojave River are in the San Bernardino Mountains (Fig. 1), where P_a exceeds 1000 mm while the Mojave Sink, more than 200 km downstream is typified by a Pa of < 80 mm. The lower reaches of the river consist of a sequence of basins; from highest to lowest these are the Manix Basin, the Cronese Basin, the Silver Lake and Soda Lake Basins at approximately equivalent elevations and, finally, Death Valley. During the late Wisconsinan and early Holocene the Silver Lake and Soda Lake Basins supported Pluvial Lake Mojave. Recent studies of sediment cores as well as lake-margin geomorphology by Wells and his colleagues (Wells *et al.*, 1987, 1989; Enzel *et al.*, 1989, 1992) have confirmed and refined the original paleohydrologic history of this lake system (Ore and Warren, 1971). The history of repeated filling and desiccation of Pluvial Lake Mojave can be used as a proxy for precipitation variations in the headwaters of the Mojave River. The relationships between anomalously wet winters in the San Bernardino Mountains, increased river flow, and inundation of the closed basins in the Mojave Sink are summarized by Enzel *et al.* (1989).

Figure 2 presents a summary of fluctuations in the level of Lake Mojave over the last 16,000 years (Wells *et al.*, 1989). The latest Wisconsinan and early Holocene history of this lake is divided into three phases. The "Intermittent Lake II" phase began with the lake's partial desiccation between 16,000 and 17,000 yr B.P. Subsequent episodes of refilling occurred during Intermittent Lake II time, punctuated by periods of drying. Between 13,000 and 14,000 yr B.P. the lake filled and did not experience partial drying again until shortly before 11,000 yr B.P. The "Lake Mojave II" phase was followed by the "Intermittent Lake III" phase that ended with final desiccation near the end of the early Holocene (Fig. 2).

An interesting aspect of the drainage of the lower Mojave River is that the Manix Basin may have been its terminus for all except the most extreme flood events until approximately 14,000 yr B.P. Meek (1989a, b) demonstrates that, shortly after 14,200 yr B.P., the sill of the

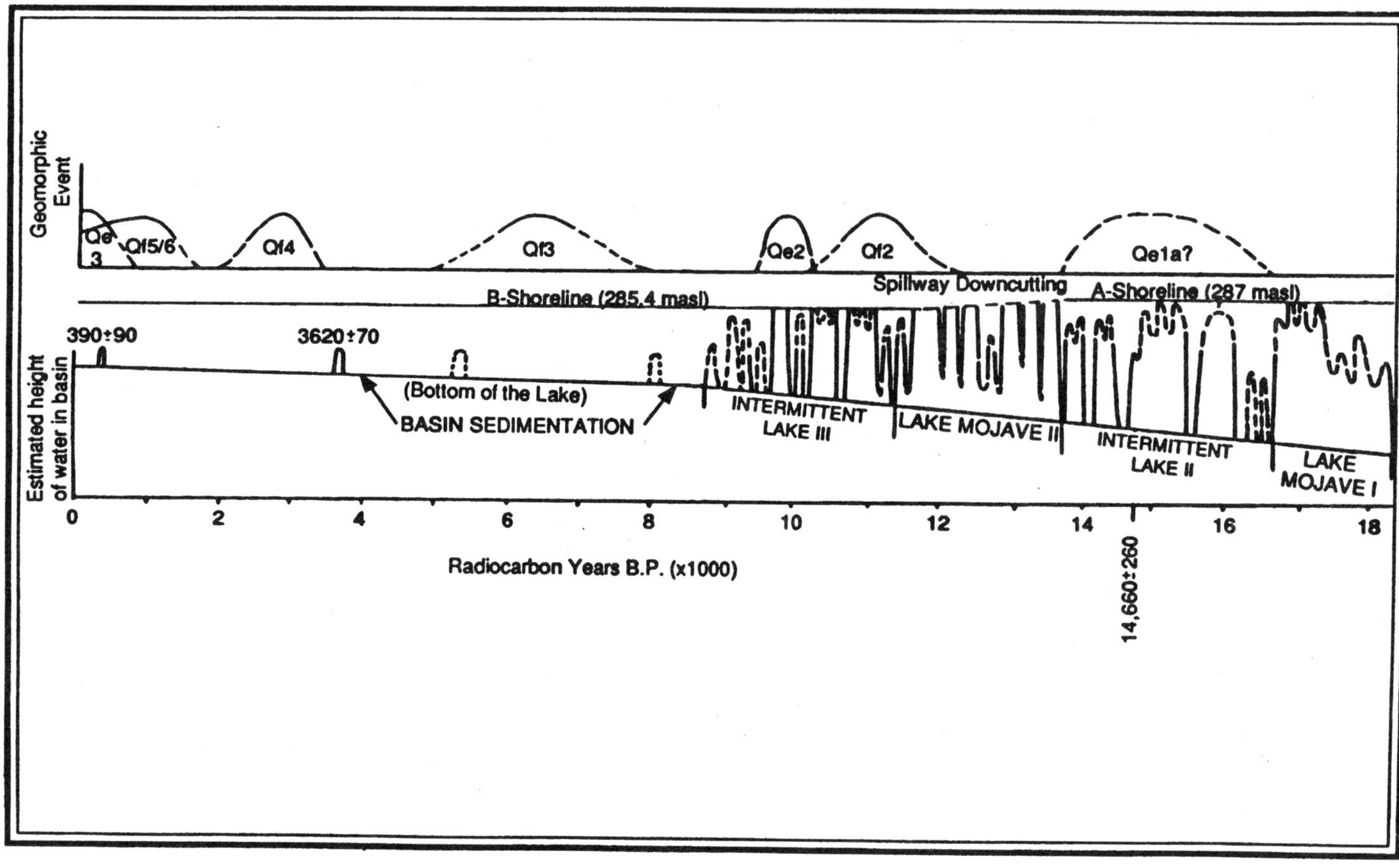

FIG. 2. Late Wisconsinan and Holocene variations in the level of Pluvial Lake Mojave, and their correlation with regional geomorphic events (from Wells *et al.*, 1989).

Pluvial Lake Manix was breached and it emptied into the Lake Mojave basin. Consequent latest Wisconsinan and Holocene down cutting in the Manix Basin resulted in the discharge of more than 3.45 km^3 of sediment into the Lake Mojave Basin. The subsequent phase of high lake stands termed Lake Mojave II, and the down cutting of the spillway connecting Lake Mojave with Lake Manly in Death Valley, may have been hydrologic responses to the input of sediment from the Manix Basin, and to the elimination of the "holding basin" represented by Lake Manix, rather than to an increase in precipitation in the headwaters of the Mojave River.

The desiccation of Lake Mojave

The timing of the final desiccation of Pluvial Lake Mojave is an important datum in the prehistory of this region. An AMS ^{14}C date of 9330 ± 95 yr B.P. on disseminated organic carbon from near the top of the last continuous lacustrine sediment interval compares well with the youngest dates on freshwater mussel (*Anodonta*) shells from two localities on the margin of the lake (9160 ± 400 and 9390 ± 140 yr B.P.; Wells *et al.,* 1989). Using a long term sedimentation rate based on two ^{14}C dates from lower in the section, the authors propose that final desiccation occurred approximately 8730 yr B.P. Although they point out that this estimated date is close to the youngest dates on subaqueous tufa reported by Ore and Warren (1971; 9160 ± 400 and 8350 ± 300 yr B.P.), Ore and Warren (*ibid.*) point out that tufa ^{14}C dates appear to be approximately 500 years too young when compared to dates on *Anodonta* shells from the same stratum. Similar conclusions were reached by Benson and Thompson (1987) in their comparison of tufa ^{14}C dates with those on packrat middens deposited above the high-water line of Pluvial Lake Lahontan. Thus, a parsimonious estimate for the timing of this event is between 8350 ± 300 and 9160 ± 400 yr B.P.

Middle(?) and late Holocene lake stands

Sediment dating to the middle and late Holocene in the subsurface cores studied by Wells *et al.* (1989) and Enzel *et al.* (1992) is composed primarily of brown, oxidized silt, sand, and clay containing sand-filled mud cracks. These playa sediments are interrupted by five lacustrine sedimentary beds, each containing ten to several hundred millimeter-scale couplets of clay and fine sand. Although no proof exists that these couplets represent annual cycles, the authors do relate these lacustrine sediments to perennial lakes that persisted for "a few years to decades." Precisely dating the two to three oldest of these middle to late Holocene lacustrine episodes is not possible due to the absence of any direct ^{14}C dates. All appear to occur stratigraphically above an erosional hiatus representing deflation of the Silver Lake playa that may date to the most arid portion of the middle Holocene (7000 to 5500 yr B. P.; Hall, 1985; Spaulding, 1991).

The timing of the youngest two Holocene lake phases in the Mojave Sink is controlled by AMS ^{14}C dates on disseminated organic carbon of 3620 ± 70 and 390 ± 90 yr B.P. (Enzel *et al.*, 1992). Evidence for the older of these two lake events consists of 350 to 400 sediment couplets, and its radiocarbon date correlates with other physical evidence for increased effective moisture in the Mojave Desert, and with the regional cooling event known as the Neoglaciation (*e.g.*, Burke and Birkeland, 1983). In the nearby Cronese Basin, Drover (1979; Fig. 1) reports a date of 3500 ± 180 yr B.P. on a lacustrine clay lens. Smith (1979) reports a date of 3520 ± 90 yr B.P. on wood from the Overburden Mud unit at Searles Lake, 120 km northwest of the Mojave Sink (Fig. 1). The Overburden Mud is related to a small (relative to Pluvial Lake Searles) lake that filled the Searles Basin during the late Holocene. Farther to the east on the Nevada-California border, Mehringer and Warren (1976) report that a period of peat formation associated with increased spring discharge in the Amargosa Desert occurred between 3980 ± 130 and 3550 ± 300 yr B.P. This is of particular interest in light of additional evidence for the sensitivity of artesian springs to precipitation variations, discussed below.

The final lake interval in the Mojave Sink, dated at 390 ± 90 yr B.P., consists of 40 to 45 sediment couplets, and correlates with the "Little Ice Age" (Enzel *et al.*, 1992), a well-known period characterized by increased glacial activity and decreased temperatures which occurred between about 550 and 100 yr B.P. (Scuderi, 1987; Grove, 1988). Mehringer and Warren (1976) record a coeval initiation of peat development in the Amargosa Desert, dated at 385 ± 120 yr B.P. Elsewhere in the Mojave Sink, large freshwater mussel middens in the Cronese Basin record the use of lake resources by prehistoric humans in what is now one of the most severe environments in North America. Eight ^{14}C dates on charcoal from these middens range from 210 ± 60 to 760 ± 50 yr B.P. (York, 1993). Charcoal from several individual "ovens" yields essentially the full range of ages. The most likely explanation for this spread of dates is that the prehistoric inhabitants were burning wood obtained from old-age mesquite (*Prosopis* sp.) trees. The charcoal from these features does include that of mesquite (York, 1993). This, however, does not exclude the possibility that there may have been more than one long-lived lake stand in the Cronese Basin over the last ca. 800 yr.

The springs of Las Vegas

When the first European explorers crossed the Mojave Desert in the Eighteenth Century, they found the bottom of the Las Vegas Valley to be a welcome oasis supporting more than a hundred artesian springs, supporting tall cottonwoods (*Populus fremontii*) that offered shade from the oppressive summer sun. In time many of the springs ceased flowing as ground water was pumped from the aquifer. In the early twentieth century paleontological investigations focused on the ancient spring sediments most extensively exposed on the north side of the Las Vegas Valley (Simpson, 1937), sediments that were ultimately named the Las Vegas Formation (Longwell *et al.*, 1965). Geochronological and paleoecological investigations of the Las Vegas Formation in the 1960s revealed that the extensive springs encountered by the first explorers were themselves mere remnants of what must have been a vast network of spring mounds, ponds, and flowing channels episodically discharging during the

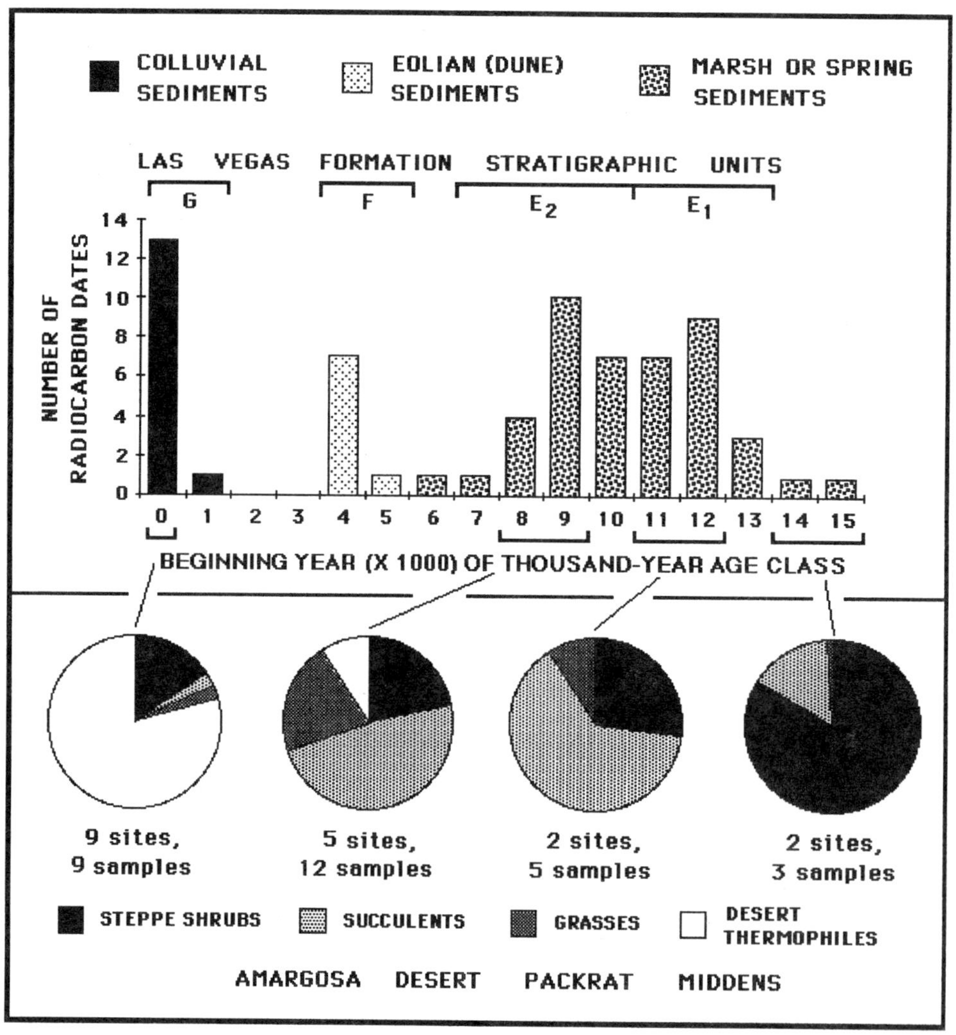

FIG. 3. Diagram comparing the age distribution and number of radiocarbon dates on late and post glacial sediments of the Las Vegas Formation (data from Haynes, 1967; Quade, 1986; Quade and Pratt, 1989; DuBarton *et al.*, 1991) with the general composition of packrat middens from below 1000 m elevation in the Amargosa Desert, ca. 100 km west-northwest of Las Vegas (from Spaulding, 1990b).

late Wisconsinan and early Holocene (Haynes, 1967; Mehringer, 1967).

Artesian springs are end-points of a hydrologic system that differs from that of a pluvial lake. High stands of a pluvial lake largely reflect runoff (*e.g.*, Enzel *et al.,* 1989), but spring discharge is dependent on aquifer recharge. The Paleozoic carbonate aquifer of the Las Vegas Valley is a confined system (Winograd and Thordarson, 1975; Quade, 1986), and increased recharge in the highlands of the Spring and Sheep Ranges (Fig. 1) would steepen the hydrologic gradient, resulting in a rapid (relative to radiocarbon-based chronological control) increase in spring discharge at the end of that gradient. Thus, episodes of enhanced precipitation in the mountains surrounding the Las Vegas Valley should be reflected in the record of discharge of the valley bottom springs.

The top portion of Figure 3 presents chronological data on younger sediments of the Las Vegas Formation, selecting only those ^{14}C dates on charcoal or disseminated organic carbon from marsh sediments. Unit E sediments are those associated with the last major episode of spring discharge in the Las Vegas Valley. Although the early explorers here found many active springs, most of those on the north side of the valley appear to have failed by about 7000 yr B.P. Unit F sediments (Fig. 3) are primarily eolian sand, relating to a major period of aridity during the middle Holocene, while Unit G is represented by colluvial silts and sands, suggesting increased erosion and deposition in arroyos during the last millennium (Haynes, 1967).

One of the most intriguing features of the Las Vegas Formation record is the absence of spring sediments that appear to date to the last glacial maximum (ca. 18,000 yr B.P.). This hiatus is marked: recent investigations in the vicinity of Tule Springs have recovered charcoal from Unit D sediments (the next oldest stratigraphic unit in the sequence) that yield dates of 26,800 ± 700 and 27,580 ± 650 yr B.P. (McVickar and Spaulding, 1993). Lack of evidence for artesian spring activity during the last glacial maximum is consistent with paleoecological data indicating a relatively dry full-glacial climate in the northern Mojave Desert. Thus, increased spring discharge during Unit E time in the Las Vegas Valley (Fig. 3) may be related to the same precipitation regime responsible for "Intermittent Lake III" in the Mojave Sink, ca. 150 km to the southwest, but there is no evidence for a correlation of spring discharge with the previous high phase ("Lake Mojave II;" Fig. 2). It is possible that storm tracks were different during the two periods. Atmospheric flow may have been dominantly zonal during Lake Mojave II time, keeping the storm tracks south of southern Nevada (Fig. 1). Stronger meridional flow may have occurred during Intermittent Lake III time, allowing storm tracks to penetrate this region. A switch from dominantly zonal to increased meridional flow is consistent with modelled changes in atmospheric circulation during deglaciation (COHMAP, 1988).

ECOSYSTEM RESPONSES: THE PALEOBOTANICAL RECORD

When did glacial-age plant communities begin changing in response to deglacial climatic change, what was the manner of those changes, and when did plant communities attain an essentially modern character? These are the questions I emphasize in considering the effects of deglacial climatic change on ecosystems. Elsewhere I have made the case that Mojavean vegetation shows early response to deglaciation (by ca. 14,000 yr B.P.), that the manner of those changes was quite complex, and that different plant communities attained essentially modern character at different times (Spaulding, 1985, 1990a,b). The impacts of subsequent postglacial climatic changes were less pronounced but still of value to understanding the final modernization of Mojavean eco-

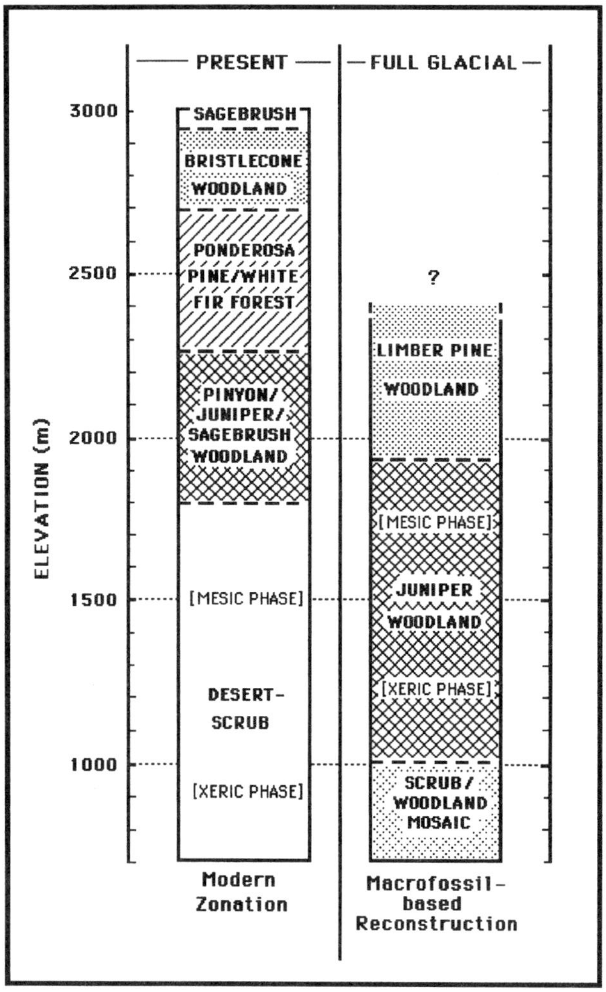

FIG 4. Diagram contrasting the present vegetation zonation in the northern Mojave Desert with that reconstructed for the last glacial maximum on the basis of packrat miden evidence from southern Nevada (modified from Spaulding, 1990c).

systems and intrinsic variations in those systems over the last ca. 8000 yr.

Initial conditions: vegetation of the last glacial maximum

Today the Mojave Desert is characterized by vast expanses of creosote bush desert scrub and, above that, blackbrush desert scrub. Neither vegetation type apparently occurred here during the last glacial maximum. Higher up on the mountains of the eastern Mojave Desert, fir-pine forest can be found today growing above pinyon-juniper woodland. The former vegetation type is also missing from our inventory of glacial-age plant communities, while there were important changes in the distribution and composition of the latter. However, the resultant differences in vegetation zonation that are summarized in Figure 4 can relay only a small part of the story.

During the last glacial maximum, between ca. 700 and 1000 m elevation, the most xeric exposures in southern Nevada supported steppe, while more mesic exposures supported xeric juniper woodland (Spaulding, 1990b). The differences are poorly understood between the steppe-shrub dominated vegetation of this region (see also Mehringer, 1967; Thompson and Mead, 1982) and the "yucca semi-desert" reported by Woodcock (1986; Wells and Woodcock, 1985) for elevations below ca. 700 m in Death Valley, less than 100 km to the west (Fig. 1). The abundance of shadscale, sagebrush, and horsebrush (*Tetradymia* spp.) in the full-glacial steppe of southern Nevada speaks to a northern affinity and suggests a relatively cold, dry climate. On the other hand, our-Lord's-candle (*Yucca whipplei*) in the vegetation of Death Valley is of a coastal California affinity, and suggests a greater increase in precipitation and mild winters (Woodcock, 1986). While the solution to this apparent conflict is not readily apparent, part of it may lie in the lower elevation of Death Valley.

Glacial-age juniper woodland continued up slope from the valley bottoms for the better part of 1000 m, with pinyon occurring only rarely, and limber pine (*Pinus flexilis*) becoming more common in the mesic phase of this woodland (Fig. 4). The abundance of limber pine here, and in subalpine woodland of the next highest vegetation zone, offers another contrast with present vegetation conditions. Today in the subalpine conifer communities of Mojave Desert mountains, limber pine is restricted relative to bristlecone pine (Spaulding, 1984; Thompson, 1990). Steppe shrubs such as shadscale, sagebrush, and horsebrush appear to have been important in both glacial-age juniper woodland, and in the lower reaches of limber pine woodland.

Deglacial environmental changes

The complex ecosystem alterations that accompanied deglacial climatic change can be viewed from either a site-specific perspective, or from a regional perspective. Each approach reveals phenomena that are of interest, just as each also reveals only part of the picture. The best examples of site-specific changes in the Mojave Desert have been discussed previously (Spaulding, 1985, 1990a), and will only be summarized here as they pertain to broader regional phenomena.

The Eleana Range-2 site, presently at the lower limit of pinyon-juniper woodland at 1810 m elevation near the northern border of the Mojave Desert (Fig. 1), supported a limber pine-steppe shrub woodland from at least 17,100 yr B.P. to between 13,200 and 11,800 yr B.P., when it gave way to pinyon-juniper woodland. Before this major turnover in dominant species there were changes in the understory. Two xerophytic shrubs, bitterbrush (*Purshia tridentata*) and dwarf goldenweed (*Haplopappus nanus*), began to increase in importance by ca. 15,000 yr B.P., long before the major turnovcr in community type occurred (Spaulding, 1985).

Based on its apparent correlation with a major resurgence to high lake levels in the Searles Basin (Smith, 1979; Fig. 1), the major vegetation turnover at the Eleana Range between 13,200 and 11,800 yr B.P. may have been due to increased temperatures accompanied by an incrcasc in precipitation near the close of the last glacial age (Spaulding, 1990b; Spaulding and Graumlich, 1986). While this hypothesis counters the conventional wisdom that the major driver of vegetation change in the Southwest at the close of the last glacial age was increased aridity, the fact that this was apparently also a time of major artesian spring activity in the Las Vegas Valley strengthens this hypothesis (Fig. 3). Further, the decline in steppe shrubs, and increase in the importance of grasses and succulents (*not* desert thermophiles) in midden assemblages from below 1000 m elevation (Fig. 3) points to milder winters but not necessarily to increased aridity.

This putative latest Wisconsinan to early Holocene pluvial event may have been due to increased monsoonal activity, in turn forced by increased solar radiation during the summer half-year (Spaulding and Graumlich, 1986; COHMAP, 1988). More recent, analog-based research (Wells *et al.*, 1989; Enzel *et al.*, 1992) supports the contention that maritime tropical air flowing into the continental interior was the source of this moisture, but also points to the likelihood that this may have occurred during the winter months as a result of disturbances similar to those associated with the modern-day El Niño effect. This is a more satisfactory paleoclimatic scenario because our understanding of the hydrology of this region shows that ground-water recharge sufficient to cause the artesian spring activation (Fig. 3) must come from winter snow melt. Summer (monsoonal) rainfall is not an effective source of ground water recharge (Winograd and Thordarson, 1975).

Differential development of plant communities

The pinyon-juniper woodland that developed at the Eleana Range site by 11,800 yr B.P. was not very different from present woodland there. At least in the mountains of the northern Mojave Desert, woodland during the waning millennia of the last glacial age and the early

Holocene bore strong resemblance to current woodland (Van Devender and Spaulding, 1979). Modernization of plant communities took millennia longer at lower elevation, desert scrub sites. The pie-diagrams presented in Figure 3 show that warm-desert shrubs were not important below ca. 1000 m elevation until after 9000 yr B.P., a delay in modernization of at least 3000 yr. These differences in the rate of modernization of ecosystems, termed differential development, are summarized in Figure 5 by measuring the taxa shared between macrofossil assemblages and the current composition of vegetation at the midden site, and arraying those values along an axis of elevation (Spaulding, 1990a). At about 18,000 yr B.P., values expressing the similarity between the modern and ancient flora at the site fell within a relatively narrow range of 20% to 35%, with one outlier at about 7%. However, the situation appears to have changed by 10,000 yr B.P. with the higher elevation records (>1700 m) showing similarity indices that range from about 50% to 70%, while the lower elevation records range from about 25% to 50% (Fig. 5).

Migrational lag was likely an important factor contributing to differential development (Cole, 1985; Ritchie, 1986). Most of the plant species that were to colonize present woodland in the Mojave Desert were in the vicinity during the last glacial maximum, just at lower elevations. Therefore, the distance that had to be crossed to arrive at their new postglacial habitats was only a few tens of kilometers (Spaulding, 1990a). Desert thermophiles such as creosote bush and white bursage, on the other hand, were displaced far to the south during the last glacial age (Cole, 1990; Van Devender, 1990); their northward migration encompassed hundreds of kilometers, and therefore took millennia longer.

Distance from full-glacial refugia, however, was not the only factor contributing to differential development. Another factor that

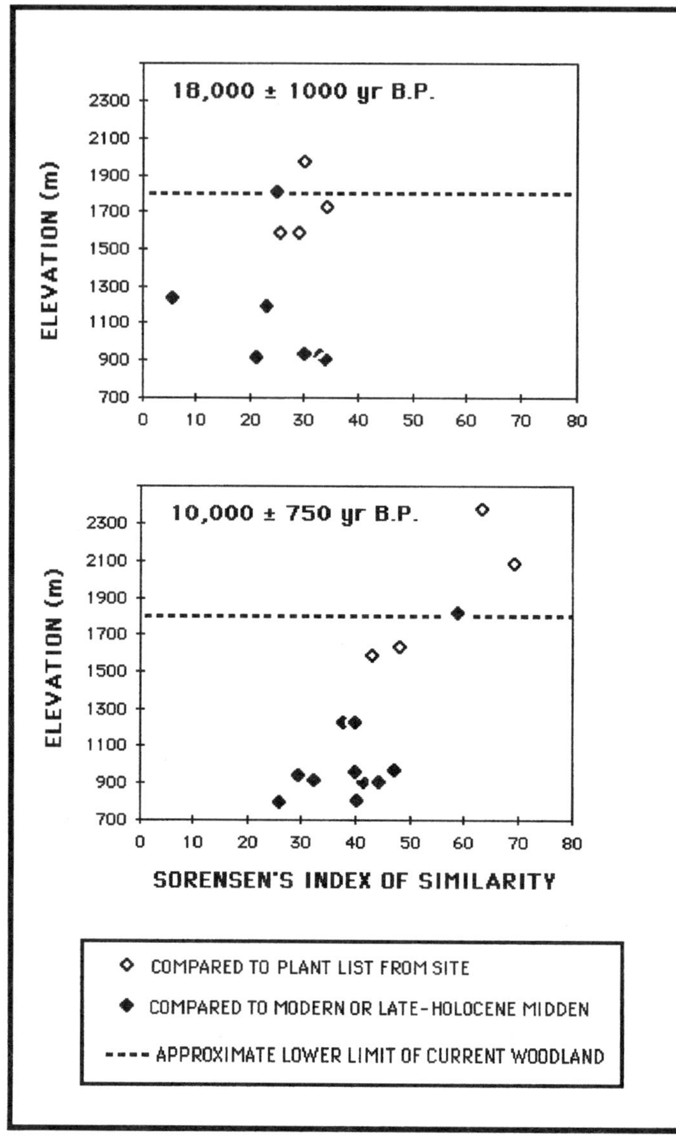

FIG 5. Indices of similarity (Sorensen, 1947) comparing macrofossil assemblages with late Holocene or modern vegetation records from the same sites, arrayed against elevation of the sites. Note the apparent trend of increasing similarity with elevation in the younger set (from Spaulding, 1990a).

may have been involved is more subtle and related not so much to the dynamics of species' migrations, but to where a fossil site lies on the local vegetation gradient. Vegetation zones are not coherent through space, but are composed of multiple overlapping species dis-

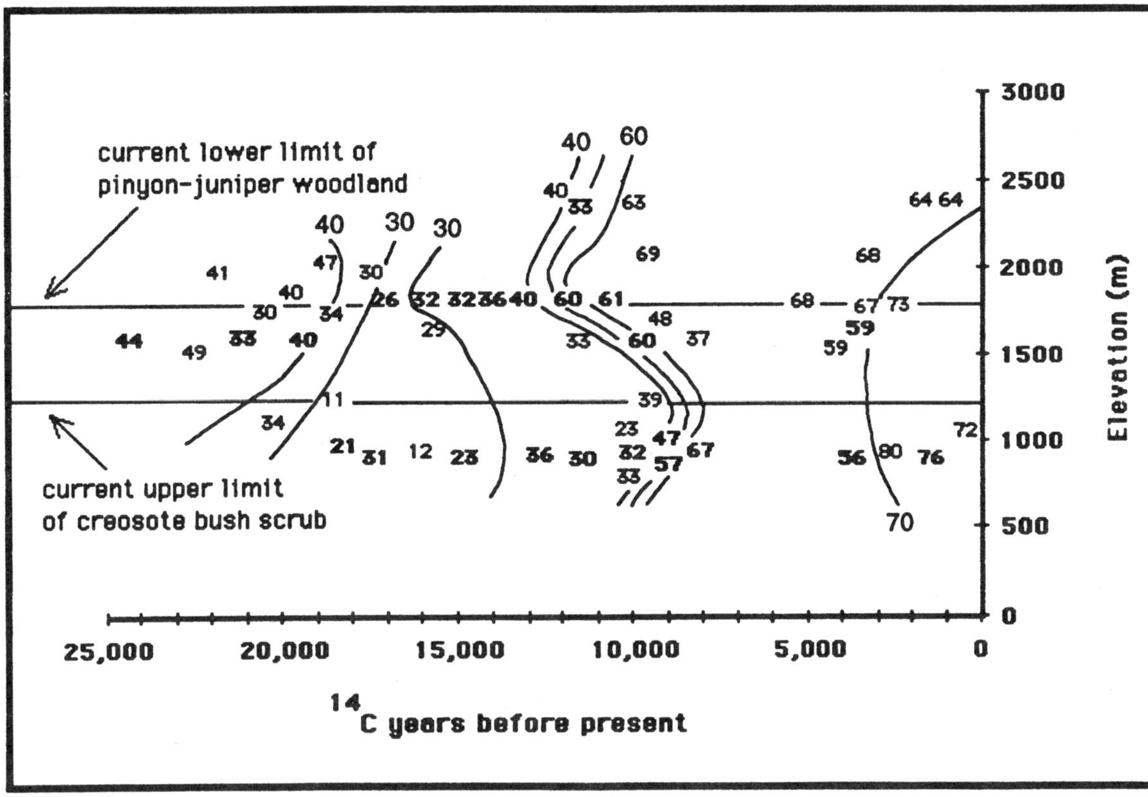

FIG 6. Diagram presenting similarity index values (bold, ancient midden assemblage compared to a modern midden from the site; plain text, ancient midden assemblage compared to a plant list from the site), and the speculative placement of isolines of similarity that may reflect the differential development of plant communities at different positions within a given vegetation zone.

tributions (*e.g.*, Whittaker, 1967). Therefore the similarity between a fossil assemblage and the modern vegetation at a site, *relative to samples higher and lower on the same gradient*, is strongly affected by its position on that gradient. In other words, modernization was rapid near the lower lower limit of pinyon-juniper woodland (Fig. 5), but it may have been slower near the upper limit of pinyon-juniper woodland where conditions were wetter and colder. Similarly, modernization was delayed between 800 and 1200 m elevation near the upper limit of current creosote bush desert scrub (Fig. 5). It could have occurred much earlier at low elevations in Death Valley. Figure 6 presents a time transgressive diagram that indicates how this phenomenon may have worked. I should note that Figure 5 presents only carefully analyzed records while Figure 6 presents all data available from the northern Mojave Desert, including samples that have not been verified since the mid-1970s. Figure 6 also shows that, even with the relatively large data set that is available, there are many gaps and the exact shape of the isolines of similarity are therefore speculative. Nevertheless, it serves to point out that the development of modern vegetation zones is a complex phenomenon (Betancourt, 1990).

Final modernization and the middle Holocene problem

The "middle Holocene problem" is actually a set of interrelated problems having to do with the nature of climate and with the availability and interpretation of paleoecological records dating to this period. Most workers agree that this was a time of higher temperatures relative to the present, but there has been sharp division regarding whether this resulted in increased aridity, or whether increased monsoonal precipitation compensated for an elevated thermal regime (compare, for example, Hall [1985] and Van Devender [1987]). This problem has been compounded by a lack of sediments and samples of the appropriate age (*e.g.*, Mehringer *et al.*, 1967; Haynes, 1967; Holliday, 1989).

It has been an exercise in frustration to attempt to record the final modernization of Mojave Desert scrub communities at the close of the early Holocene, or during the first millennium of the middle Holocene, because we do not know exactly when the hallmark of warm-desert scrub, creosote bush, became established over much of its range. Many other warm-desert shrubs arrived in the southern and central Mojave Desert before creosote bush, and its migration appears to have been particularly slow relative to such species as white bursage, brittle bush, and desert spruce. It appeared near the southern boundary of the Mojave in the Marble Mountains by 7900 yr B.P., and near its present northern limit in the Eureka Valley by 3900 yr B.P. (Spaulding, 1990b). Throughout most of creosote bush's broad range in the Mojave Desert, however, we can only find packrat middens that are older than ca. 8000 yr B.P. and therefore lack this

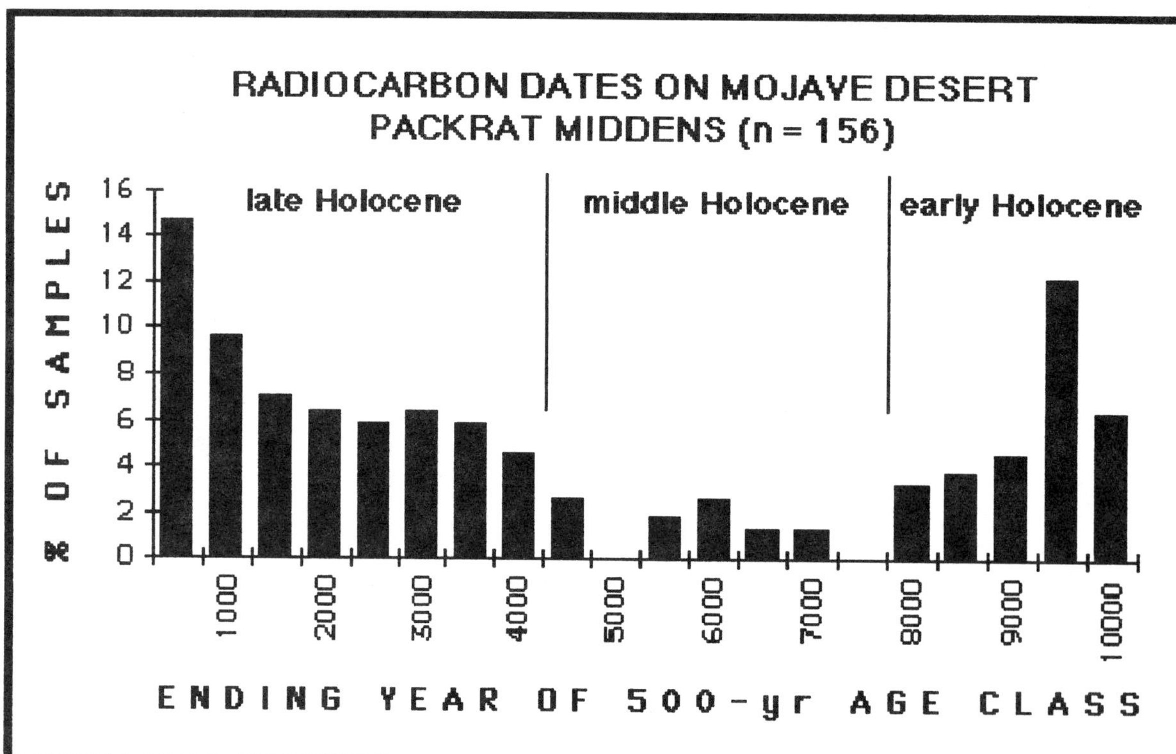

FIG. 7. The distribution of radiocarbon dates on middens, in 500-year age classes, from the Mojave Desert.

shrub, or middens that are younger than ca. 5500 yr B.P., in which case creosote bush is common to abundant.

The difficulty of finding packrat midden records that spanned the middle Holocene led to an assessment of the temporal distribution of middens from the Mojave Desert. A simple histogram of ^{14}C dates on middens, which is taken as a proxy of midden abundance over the last 10,000 yr, shows that middle Holocene middens are indeed rare (Fig. 7). Rather than being an artifact of researcher bias (Webb and Betancourt, 1990), it is more likely that the general dearth of middens dating to this period represents a real phenomenon—a drastic decline in packrat population density due to increased aridity and, therefore, the production of fewer middens. The hypothesized relationship between effective moisture (or lack thereof) and the number of middens recovered from a discrete area is outlined in Figure 8. The dependence of rodent populations on primary productivity in arid to semi-arid regions is well known, as is the direct correlation between primary productivity and rainfall (Beatley, 1976; Rosenzweig, 1968; Shmida *et al.*, 1986). Increased aridity leading to reduced populations of *Neotoma* may explain why middle Holocene middens are so scarce in many parts of the Southwest (Fig. 7).

Although the concept of an arid middle Holocene is disputed by some authors (*e.g.,* Van Devender *et al.*, 1987), widespread packrat, pollen, and geomorphic evidence indicates that the period from ca. 7500 to 4500 yr B.P. was, as Antevs (1952) had originally suggested, particularly arid throughout most of the Southwest (*e.g.*, Byrne *et al.*, 1979; Hall, 1985; Holliday, 1989; Mehringer, 1967; Spaulding, 1991). It would appear that the only areas that enjoyed increased effective moisture during the middle Holocene were the Colorado Plateau (Betancourt, 1990; Petersen, 1988), and the mountains (but not the valleys [Mehringer, 1986]) of the central and eastern Great Basin (Thompson, 1990). It therefore seems reasonable to speculate that creosote bush's advance through much of the Mojave Desert may have been enhanced during the middle Holocene as a consequence of the competitive advantage it would have gained over other species during a period that was more arid than the present.

Ecosystem responses to late Holocene climatic change

The accurate dating of late Holocene lacustrine events in the Mojave Sink provides an opportunity to test for corresponding changes in nearby desert ecosystems using packrat midden records. One such record has recently been developed for the Neoglaciation in the Little Granite Mountains of Fort Irwin (as opposed to the better-known Granite Mountains south of the Kelso Dunes; Spaulding, 1993). Two others have been developed for the Little Ice Age, by Cole and Webb (1985) for the Greenwater Valley, California, and by Hunter (1991) for desert scrub in the vicinity of Searchlight, Nevada (Fig. 1).

The Neoglaciation

Four packrat middens recovered from the Little Granite Mountains lie at 1320 to 1360 m elevation in mixed desert scrub. Creosote bush scrub occurs nearby at slightly lower elevations, and blackbrush scrub occurs on mesic slopes and at slightly higher elevations. Common plants in the vicinity include goldenweed (*Haplopappus linearifolius*), California buckwheat (*Eriogonum fasciculatum*), desert almond, and joint-fir (*Ephedra viridis*). Less common mesophytes that occur in the vicinity and provide clues to changes in effective moisture during the late Holocene include wedgeleaf goldenweed (*Haplopappus cuneatus*) and blackbrush. With nine macrofossil assemblages ranging in age from 700 ± 60 to 11,470 ± 70 yr B.P., the Little Granite Mountains middens pro-

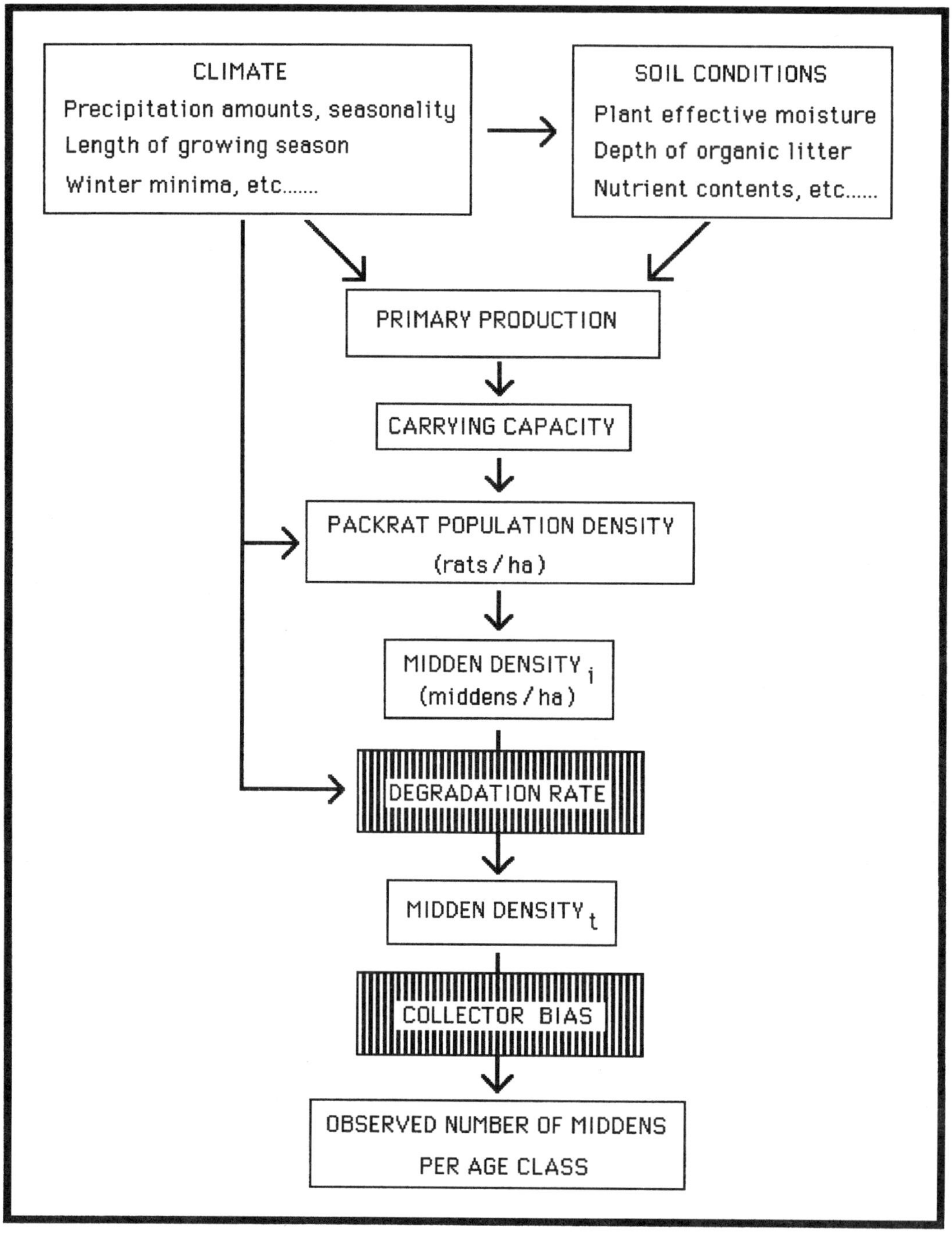

FIG. 8. Flow chart showing a hypothetical relationship between climate and the number of packrat middens available in a given area.

vide a detailed record of vegetation change in the central Mojave Desert.

For the sake of brevity this discussion of the Little Granite Mountains (LGM) record is limited to six macrofossil assemblages from the LGM-2 and -4 sites that reflect conditions during and after the Neoglaciation. Each fossil record is controlled by a modern midden from the same rock shelter. I further focus on those three mesophytic plant species that typically occupy more mesic desert scrub at elevations above *ca.* 1300 m and may be the most sensitive indicators of variations in effective moisture: blackbrush, wedge-leaf goldenweed, and Mojave sage. Mojave sage is taken to be the most sensitive indicator of climate change because it is absent from the locality today. It is also most abundant in the two oldest late Holocene assemblages (Fig. 9) and, therefore, maximum effective moisture (relative to the present and within the time span sampled) appears to have occurred between 3160 and 3490 yr B.P. The decline in the abundance of Mojave sage after that time suggests increased aridity, although the relative abundance of wedge-leaf goldenweed and blackbrush in two younger assemblages from the site suggests that levels of effective moisture above those of the present continued until after 2640 yr B.P. (Fig. 9).

The midden samples from the LGM-2 site record vegetation at a locality somewhat more mesic than the LGM-4 site. Blackbrush is an important component of the modern midden there, while it is essentially absent from the modern midden at the LGM-4 site (Figs. 9, 10). Therefore, the LGM-2 assemblages, both dated at ca. 1450 yr B.P., indicate that conditions were somewhat more arid than those of the present

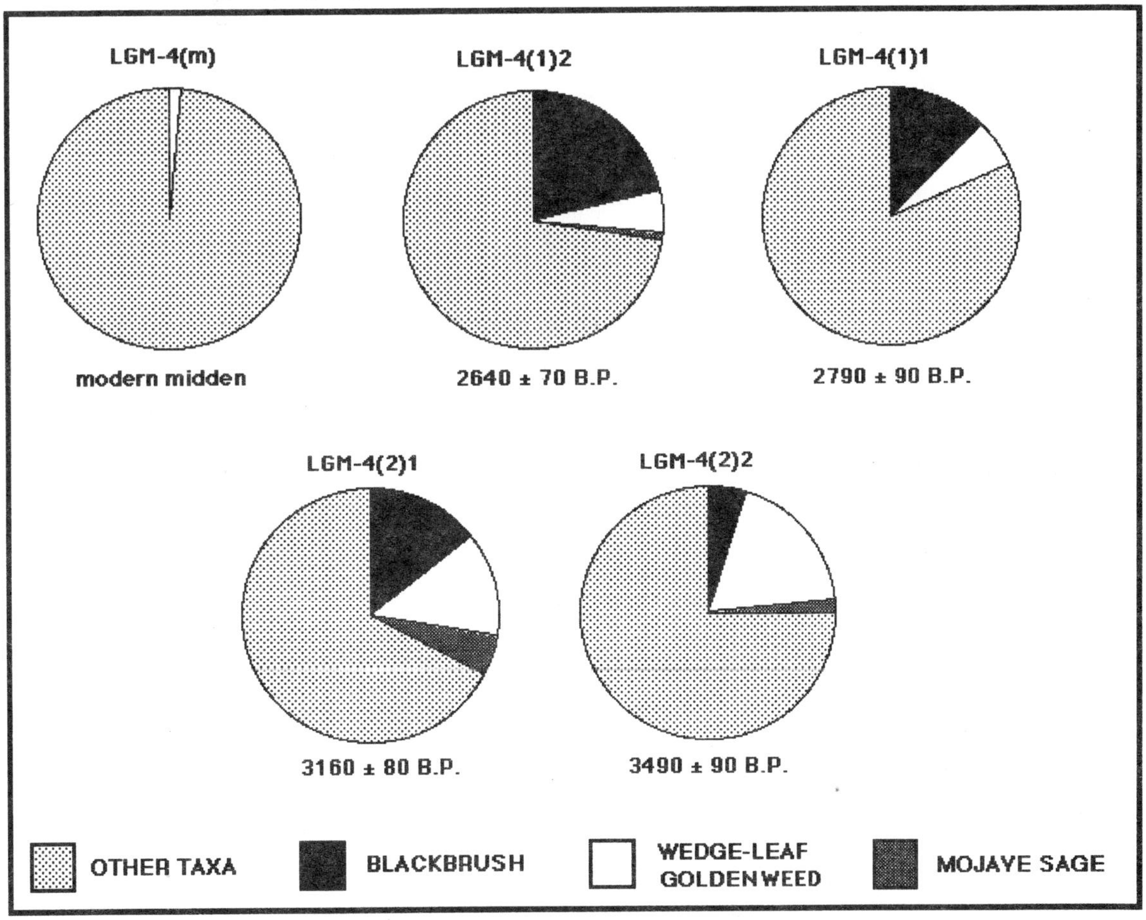

FIG. 9. Pie diagrams showing changes in the relative abundance of indicator species in macrofossil assemblages from the Little Granite Mountains-4 packrat midden site in the central Mojave Desert.

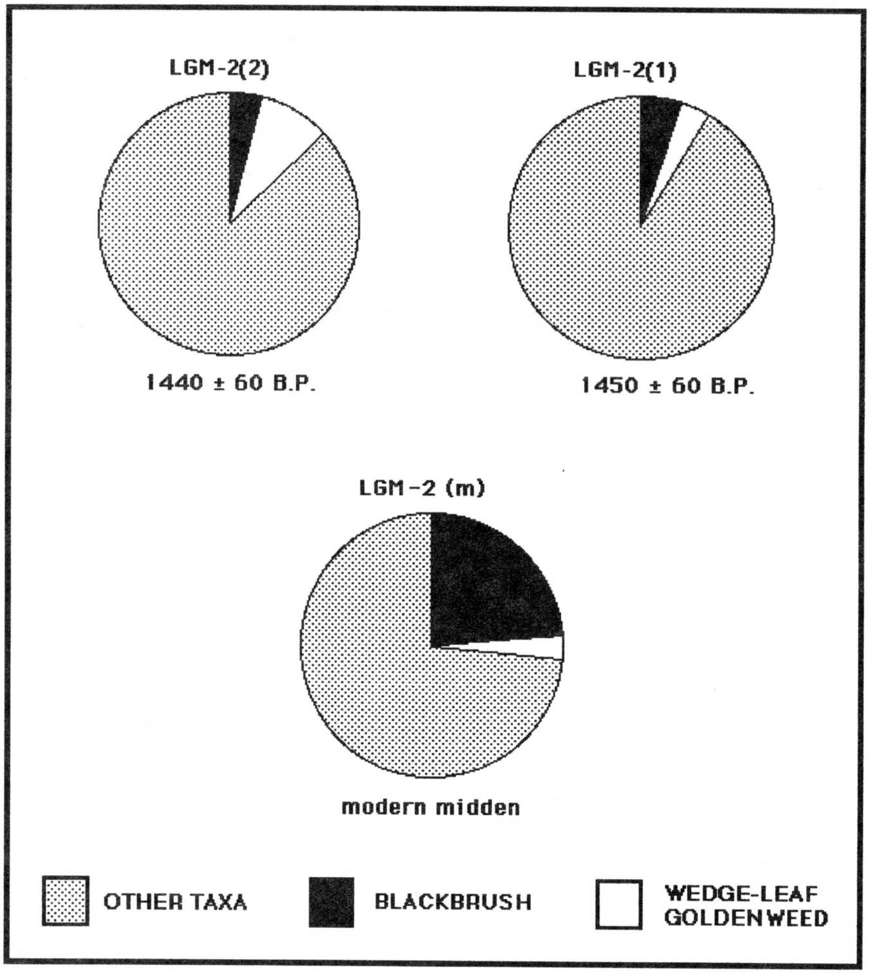

FIG. 10. Pie diagrams showing changes in the relative abundance of indicator species in macrofossil assemblages from the Little Granite Mountains-2 packrat midden site in the Central Mojave Desert.

due to the smaller amount of blackbrush in those samples (Fig. 10). The records from this site constrain (albeit loosely) the end of the Neoglacial period of increased effective moisture to between ca. 2650 and ca. 1450 yr B.P. Middens from farther north on the eastern margin of the Mojave Desert at Fortymile Canyon (Fig. 1) record an upward advance in the upper limit of creosote bush, and therefore warmer winters, beginning by ca. 1470 yr B.P. (Spaulding, 1990b).

These data suggest that, while the Neoglacial lake in the Mojave Sink, dated at 3620 ± 70 (Enzel *et al.*, 1992), may have persisted as long as 400 yr based on the number of sediment couplets preserved in that section of core, mesic vegetation persisted considerably longer in the nearby mountains (Fig. 9). Comparison of the vegetational record from the LGM-4 site with the paleohydrologic record from the Mojave Sink therefore indicates that, while enhanced precipitation in the San Bernadino Mountains may be the primary cause of lake filling (Wells *et al.*, 1989), it was not restricted to the headwaters of the Mojave River. The lacustrine interval at about 3600 yr B.P. likely represents merely a peak event(s) recorded by a system with a high threshold, an observation consistent with the analyses of Enzel *et al.* (1992).

The Little Ice Age

The global cooling event known as the Little Ice Age began between 750 and 500 yr B.P., and ended during the nineteenth century (Grove, 1988). Studies in the southern Sierra Nevada show that here, as elsewhere, this period was characterized by a series of glacier advances and retreats (Scuderi, 1987). At springs in the Amargosa Desert, Mehringer and Warren (1976) record a renewed period of peat formation and possible spring activity dated at 385 ± 120 yr B.P. During the same period there appears to have been one long-lived perennial lake in the Mojave Sink, dated to 360 ± 60 yr B.P. (Enzel *et al.*, 1992). Two midden series from Mojave Desert provide perspectives on the responses of desert ecosystems to Little Ice Age climatic change. The first, from the Greenwater Valley, ca. 170 km north of the Mojave Sink, is a detailed study of changes at the lower limit of blackbrush desert scrub (Cole and Webb, 1985). The second is an overview of evidence for similar changes in the desert near Searchlight, Nevada (Hunter, 1991; Fig. 1).

Cole and Webb (1985) show that the descent of blackbrush into the elevational range occupied by creosote bush, from ca. 1410 m to 1340 m elevation, occurred between 290 ± 50 and 270 ± 70 yr B.P., perhaps more than 100 radiocarbon years *after* the Little Ice Age lake stand recorded in the Mojave Sink. However, another mesophyte in this record, California buckwheat, appears to display an earlier response. It descended to a new and lower elevation between 1340 ± 140 and 290 ± 50 yr B.P. (Cole and Webb, 1985: table 2). The higher sensitivity to climatic change of California buckwheat relative to blackbrush is indicated further by the absence of the former shrub from most modern middens at the study site today, and the persistence of the latter. In other words, this record suggests that California buckwheat may respond more quickly to changes in effective moisture, while blackbrush displays a higher degree of inertia.

Farther south in the vicinity of Searchlight, middens recovered by Hunter (1991) suggest a different chronology of vegetation responses to Little Ice Age climatic change. Blackbrush first appeared in sites between 1287 and 1220 m elevation between 840 ± 60 and 600 ± 50 yr B.P. Establishment of a plant community in which blackbrush was at least common occurred by 470 ± 60 yr B.P. The earlier record for blackbrush displacement at this locality, relative to the more northerly Greenwater Valley sites, is intriguing. Being lower in elevation one might expect a later response to increased effective moisture during the Little Ice Age. Because the response occurred earlier, factors must be involved other than site elevation. It is possible that increased winter precipitation during this period (Cole and Webb, 1985) had greater effect in the vicinity of Searchlight because this locality lies farther south, and is less affected by the rain shadow of the Sierra Nevada and larger mountain ranges of the Mojave Desert. The Greenwater Valley site, on the other hand, lies in the lee of not only the Sierra Nevada, but also the Panamint Range (Fig. 1).

THE FATE OF MOJAVEAN FAUNA

Prior to megafaunal extinctions at the close of the Wisconsinan glacial age, the ecosystems of the Mojave Desert apparently had a fully stocked second trophic level. Components of these ecosystems that are most notably absent today are the large, extinct herbivores, including camel (*Camelops hesternus*), llama (*Hemiauchenia* sp.), native American horses (*Equus occidentalis*, *Equus* subgen. *Asinus*), mammoth (*Mammuthus columbi*), ground sloths (*Nothrotheriops shastensis*, *Megalonyx* sp., *Glossotherium* sp.), and bison (*Bison* sp.). The camelids, equids, and mammoth are generally well-represented at fossil localities throughout the region, while the ground sloths (with the exception of the Shasta ground sloth) and bison are rare. Mastodon (*Mammut* sp.) occurs in older deposits of the Irvingtonian Land Mammal Age, but appears to have been restricted to coastal California, and to regions east of the Great Basin and Sonoran Desert during the last glacial age (Jefferson, 1989, 1991; Mawby, 1967; Miller, 1987). Smaller mammals that occurred in the Mojave Desert during the last glacial age, and that survive elsewhere today, include the sagebrush vole (*Lagurus curtatus*), marmot (*Marmota flaviventris*), and pika (*Ochotona princeps*). The marmot and pika appear to have been relatively common in the mountain ranges of the Mojave Desert during the Wisconsinan (*e.g.*, Mead and Spaulding, this volume; Mehringer and Fergusson, 1969; Reynolds *et al.,* 1991 a, b).

The large carnivores of glacial-age ecosystems are much more poorly represented in the fossil record. Their remains can be expected to be rare relative to their prey, but the general dearth of large, extinct carnivores in the Mojave Desert fossil record is intriguing. The American lion (*Felis atrox*) is reliably recorded for one site (Jefferson, 1991). There are no records of the dire wolf (*Canis dirus*), short-faced bear (*Arctodus simus*), or saber-tooth cat (*Smilodon* sp.). This is highlighted further by the apparent association between cave sites and the remains of dire wolf in the rest of North America (Kurten and Anderson, 1980). Caves yielding the remains of late Pleistocene fauna are relatively common in the Mojave Desert, but dire wolf fossils remain elusive. There are several records of *Felis* sp. that could be referable to the mountain lion (*F. concolor*), to the American lion, or to the American chetah (*Acionyx* sp.). On the other hand, smaller carnivores and scavengers such as the coyote (*Canis latrans*) and vultures (*e.g.*, *Teratornis merriami*, *Gymnogyps californianus*) appear to be relatively widely represented.

In considering the faunal changes that occurred at the close of the last glacial age it is appropriate, in this forum at least, to consider two radically different mechanisms: climatic change and predation by the first humans to enter this continent (see contributions in Martin and Klein, 1984). The restriction of sagebrush vole, marmot, and pika to higher latitudes and elevations during deglaciation, and their consequent local extinction in the Mojave Desert highlands, probably was forced by climatic change. For several reasons, best articulated by Martin (1984, 1986), the same argument cannot be easily made for the terminal extinction (as opposed to local extinctions) of the megafauna that inhabited this region. As a working hypothesis I adopt the notion that both climatic change *and* human predation by highly specialized hunters (*e.g.* Martin, 1973) had an affect on the fauna of this region, and that the latter should be (at the very least) seriously entertained if the former conflicts with the existing paleoenvironmental record. Deglacial climatic changes forced alterations in the distribution of boreal mammals (*e.g.*, Grayson, 1982) and, as discussed below, also affected the Shasta ground sloth *prior* to its final extinction at ca. 10,900 yr B.P. It remains to be seen whether a climatic explanation for the terminal extinctions of large mammalian taxa can be integrated into the framework provided by the paleoenvironmental record. The main objective here is to point to some of the ecological and biogeographic phenomena that need to be integrated into a holistic explanation of climatically forced extinction.

Extinctions and environmental change

We know a great deal about the Late Quaternary paleoecology of three herbivores of the glacial-age Mojave Desert: the packrat, the pika, and the Shasta ground sloth. Others deserve mention, but these three taxa illustrate many of the fundamental contrasts in ecological and evolutionary history that are germane to this forum. The "packrat" is represented by many species of *Neotoma* and, moreover, appears to have been characterized by measurable evolutionary plasticity during the Wisconsinan (*e.g.*, Harris, 1984). The pika, on the other hand, is an evolutionarily conservative genus currently represented by only two species in the Western Hemisphere, one of which (*Ochotona princeps*) is widely distributed in the West. Skeletal material and dung pellets of *O. princeps* occurs in packrat middens recovered from Great Basin and Mojave Desert mountain ranges, south of its current interior distribution (see Mead and Spaulding, this volume). Finally, the Shasta ground sloth was a monospecific genus, based on morphological features that distinguish *Nothrotheriops shastensis* from the South American ground sloth *Nothrotherium*, and from North American species of *Glossotherium* such as Harlan's ground sloth (*G. harlani*; Paula Cuoto [1971] cited in Kurten and Anderson, 1980).

Establishing which of these taxa are monotypic within the time span and area considered is important to a consideration of their niche and habitat breadth in glacial-age ecosystems. The pika and the Shasta ground sloth appear to be phenotypically conservative and, therefore, we are reasonably certain that the range of habitats they occupied during the last glacial age, and the range of diets demonstrated by analyses of their dung, are indicative of the ecological plasticity of a single species.

The habitat breadth of the Shasta ground sloth has been discussed elsewhere, as has its niche breadth inferred from the analyses of its mummified dung (Hansen, 1978; Spaulding and Martin, 1979; Thompson *et al*., 1980). Central features of these findings are that this bear-sized herbivore subsisted on plant taxa that are presently common throughout the desert West (*e.g*., *Yucca* sp., *Ephedra* sp., *Sphaeralcea* sp., *Opuntia* sp.) and shunned taxa that were adversely affected by deglacial climatic changes in the desert (*e.g*., *Juniperus* spp., *Pinus* sp.), and that it appears to have occupied a wide range of habitats, at least on a seasonal basis, during the closing millennia of the Late Wisconsinan. Ground sloth dung balls have been recovered from caves as high as 2000 m, to those as low as 600 m elevation (*e.g*., Van Devender *et al*., 1979; Long *et al*., 1974). Significantly, though, there is only one record of the Shasta ground sloth dating to the full glacial (ca. 18,000 yr B.P.) in all of western North America. A possible exception is from the lowest elevation site, Gypsum Cave (21,470 ± 760 yr B.P.; Mead [1979] cited in Thompson *et al.* [1980]). In the well-dated stratigraphic sequence from Rampart Cave (Fig. 1), for example, ground sloth occupation during the middle Wisconsinan interstadial was followed by the apparent abandonment of the cave during the full glacial. Subsequent reoccupation of this site occurred by 14,000 yr B.P., and occupation continued until their extinction ca. 10,900 yr B.P. (Long *et al*., 1974). It is likely that this Neotropical species was excluded from much of the region by cold winter temperatures during the last glacial maximum (Spaulding, 1990b, p. 196). Its reoccupation of Rampart Cave is a reflection of its northward spread in response to warmer winters.

Pika, on the other hand, appear to have been common during the last glacial maximum. In the northern Mojave Desert packrat middens yielding pika remains range in age from >44,600 to ca. 14,300 yr B.P. (Mead and Spaulding, this volume). Their apparent absence from the low elevation (<2000 m) fossil record after ca. 14,300 yr B.P., and the appearance of the Shasta ground sloth at Rampart Cave at about the same time, are both likely related to warming at the end of the last glacial age. That the period between ca. 15,000 and 12,000 yr B.P. was marked by significant warming is demonstrated by both global and regional records (*e.g*., Mayewski *et al*., 1993; Mix, 1987; Spaulding, 1990b). It is logical that the northward spread of the Neotropical Shasta ground sloth would have been largely coincident with the northward restriction of a species with boreal affinities such as the pika.

So is there any intrinsic aspect of these records that might accommodate a climatic explanation for both Shasta ground sloth terminal extinction, and the local extinctions and northward restriction of pika? There is none to my eye. Indeed, as it is said in Las Vegas, the "smart money" would have been on the terminal extinction of pika not the ground sloth. The former is a boreal taxon restricted to relatively mesic habitats (*e.g.*, Hafner, 1993) and by warm summers (Mead and Spaulding, this volume). It would have been ill-suited to the post-glacial environment. The ground sloth, with its preference for desert shrubs and succulents, would have been expected to continue in its northward spread into the Mojave Desert as desert ecosystems became more widespread. Climatic explanations for terminal extinctions seem to fit poorly the actual paleoclimatic and paleoecological record.

SOME AFTERTHOUGHTS

In this paper I have attempted to summarize some of the more intriguing features of the paleoenvironmental record of the Mojave Desert. The loose framework of this presentation has been structured according to the phenomena considered: paleohydrologic records, paleobotanical records and, finally, paleontologic records. The physical environment first, then the vegetation and, finally, some consideration of the fauna. Throughout, the demonstration of correlations sufficiently tight to warrant a discussion of causal interactions between these systems is crucial. Although far from ideal, the chronologies of several Mojave Desert records are sufficiently well-controlled to make this a worthwhile undertaking.

The paleohydrologic studies from the Mojave Sink and the springs of southern Nevada provide records of runoff and discharge, which are some of the simplest proxy climate records of precipitation variations. Hopefully there will be further attempts to use these as baselines from which to compare the nature and timing of vegetation responses to climatic change. Careful consideration of the location of drainage basins and recharge areas is required to assure that the same regional climatic changes are being measured by, for example, both a pluvial lake record and the middens located in the highlands nearby. This paper has dealt with phenomena that lie within the latitudinal bounds of the Mojave Desert, and therefore we can be fairly certain that we are viewing different systems responding to the same climatic changes. This established, we then have the fascinating task of studying the responses and inferring causal mechanisms.

Even with a relatively large packrat midden database, the dynamic aspects of deglacial vegetation change in the Mojave Desert are just beginning to reveal themselves. This level of understanding obviously must be raised further if we are to develop a "grand unified theory" that would explain the extinctions of megafauna at the end of the last glacial age in terms of environmental change. We can only hope to do so by accumulating much more data on the paleoclimate and paleovegetation of the periods before, during, and after the extinction event, and on the faunas that were affected by deglacial environmental change. Attempts at falsification not withstanding (Martin, 1986), the "overkill" theory will likely stand as a viable hypothesis accounting for terminal Pleistocene extinctions until such time as a comprehensive environmental hypothesis can be forwarded.

ACKNOWLEDGMENTS

Particular thanks go to Dames & Moore, Inc., and the Quaternary Research Center and Department of Botany, University of Washington, Seattle, to Mifflin and Associates of Las Vegas, Nevada, and to the U. S. Geological Survey and the National Science Foundation, for support in various phases of this research. M. Elizabeth Manian of B & G Environmental Services kindly provided Figure 1. And, finally, I wish to thank a Teacher I met on my road, who suggested that I treat "why" questions with particular care.

REFERENCES

Anderson, R. S. (1990). Development of Southwestern ponderosa pine forests: what do we really know? *U. S. D. A. Rocky Mountain Forest and Range Experiment Station, General Technical Report* RM185, 15-22.

Andrews, J. T., Carrara, P. E., King, F. B., and Stuckenrath, R. (1975). Holocene environmental changes in the alpine zone, northern San Juan Mountains, Colorado: evidence from bog stratigraphy and palynology. *Quaternary Research* **5**, 173-197.

Antevs, E. (1952). Cenozoic climates of the Great Basin. *Geologisches Rundschau* **40**, 94-108.

Barnosky, C. W., Anderson, P. M., and Bartlein, P. J. (1987). The northwestern U.S. during deglaciation: vegetation history and paleoclimatic implications. *In* "The Geology of North America, Volume K-3, North America and Adjacent Oceans During the Last Deglaciation" (W. F. Ruddiman and H. E. Wright, Jr., Eds.), pp. 289-321. Geological Society of America, Boulder, CO.

Beatley, J. C. (1975). Climates and vegetation pattern across the Mojave/Great Basin Desert transition of southern Nevada. *American Midland Naturalist* **93**, 53-70.

Beatley, J. C. (1976). Rainfall and fluctuating plant populations in relation to distributions and numbers of desert rodents in southern Nevada. *Oecologia* **24**, 21-42.

Benson, L. V., and Thompson, R. S. (1987). Lake-level variation in the Lahontan Basin for the past 50,000 years. *Quaternary Research* **28**, 69-85.

Betancourt, J. L. (1990). Late Quaternary biogeography of the Colorado Plateau. *In* "Fossil Packrat Middens: The Last 40,000 Years of Biotic Change" (J. L. Betancourt, T. R. Van Devender, and P. S. Martin, Eds.), pp. 259-292. University of Arizona Press, Tucson.

Brown, D. E., Lowe, C. H., and Pase, C. P. (1979). A digitized classification system for the biotic communities of North America, with community (series) and association examples for the Southwest. *Journal of the Arizona-Nevada Academy of Science* **14**, Supplement 1.

Burke, R. M., and Birkeland, P. W. (1983). Holocene glaciation in the mountain ranges of the western United States. *In* "The Holocene" (H. E. Wright, Jr., Ed.), pp. 3-11. University of Minnesota Press, Minneapolis.

Byrne, R., Busby, C., and Heizer, R. F. (1979). The Altithermal revisited: pollen evidence from Leonard Rockshelter. *Journal of California and Great Basin Anthropology* **1**, 280-294.

CLIMAP Project Members (1981). Seasonal reconstructions of the earth's surface at the last glacial maximum. *Geological Society of America Map and Chart Series* **MC-36**.

Cole, K. L. (1985). Past rates of change, species richness, and a model of vegetational inertia in the Grand Canyon, Arizona. *American Naturalist* **125**, 289-303.

Cole, K.L. (1990). Reconstruction of past desert vegetation along the Colorado River using packrat middens. *Palaeogeography, Palaeoclimatology, Palaeoecology* **76**, 349-366.

Cole, K.L., and Webb, R.H. (1985). Late Holocene vegetation changes in Greenwater Valley, Mojave Desert, California. *Quaternary Research* **23**, 227-235.

COHMAP Project Members (1988). Climatic changes of the last 18,000 years: observations and model simulations. *Science* **241**, 1043-1052.

Drover, C. E. (1979)."The Late Prehistoric Human Ecology of the Northern Mohave Sink San Bernardino County, California". Unpublished Ph. D. dissertation, University of California, Riverside.

DuBarton, A. E., Spaulding, W. G., Kelly, M. S., and Cleland, J. H. (1991). "City of Las Vegas Land Transfer Final Report on Archaeological and Paleoenvironmental Testing of Three Sites Along the Eglington Escarpment." Dames & Moore, Las Vegas, Nevada.

Enzel, Y., Cayan, D. R., Anderson, R. Y., and Wells, S. G. (1989). Atmospheric circulation during Holocene lake stands in the Mojave Desert: evidence of regional climate change. *Nature* **341**, 44-48.

Enzel, Y., Brown, W. J., Anderson, R. Y., McFadden, L. D., and Wells, S. G. (1992). Short-duration Holocene lakes in the Mojave River drainage basin, southern California. *Quaternary Research* **38**, 60-73.

Grayson, D. K. (1982). Toward a history of Great Basin mammals during the last 15,000 years. *In* "Man and Environment in the Great Basin" (D. B. Madsen and J. F. O'Connel, Eds.), pp. 82-101. *Society for American Archaeology Papers* **2**.

Grove, J. M. (1988). "The Little Ice Age." Methuen, New York.

Hafner, D. J. (1993). North American pika (*Ochotona princeps*) as a Late Quaternary biogeographic indicator species. *Quaternary Research* **39**, 373-380.

Hall, S. A. (1985). Quaternary pollen analysis and vegetational history of the Southwest. *In* "Pollen Records of late Quaternary North American Sediments" (V. M. Bryant, Jr., and R. G. Holloway, Eds.), pp. 95-124. American Association of Stratigraphic Palynologists Foundation, Dallas, Texas.

Hansen, R.M. (1978). Shasta ground sloth food habits, Rampart Cave, Arizona. *Paleobiology* **4**, 302-319.

Harris, A. H. (1984). *Neotoma* in the late Pleistocene of New Mexico and Chihuahua. *In* "Contributions in Quaternary Vertebrate Paleontology" (H. H. Genoways and M. R. Dawson, Eds.). *Carnegie Museum of Natural History Special Publication* **8**, 164-178.

Haynes, C. V., Jr. (1967). Quaternary geology of the Tule Springs area, Clark County, Nevada. *In* "Pleistocene Studies in Southern Nevada" (H.M. Wormington and D. Ellis, Eds.). *Nevada State Museum Anthropological Papers* **13**, 15-104.

Holliday, V. T. (1989). Middle Holocene drought on the southern High Plains. *Quaternary Research* **31**, 74-82.

Hopkins, D. M., (1975). Time-stratigraphic nomenclature for the Holocene epoch. *Geology* **3**, 10.

Hunter, K. L. (1991). "Impacts of the Little Ice Age on the Vegetation of Southern Nevada." Unpublished M.S. thesis, University of Nevada, Las Vegas.

Huntley, B, and Webb, T., III. (1989). Migration: species' response to climatic variations caused by changes in the Earth's orbit. *Journal of Biogeography* **16**, 5-19.

Jefferson, G. T. (1989). Late Pleistocene and earliest Holocene fossil localities and vertebrate taxa from the western Mojave Desert. *In* "The West-Central Mojave Desert: Quaternary Studies Between Kramer and Afton Canyon" (R. E. Reynolds, Ed.), pp. 27-40. San Bernardino County Museum Association Special Publication, San Bernardino, CA.

Jefferson, G. T. (1991). Rancholabrean age vertebrates from the southeastern Mojave Desert, California. *In* "Crossing the Borders: Quaternary Studies in Eastern California and Southwestern Nevada" (R. E. Reynolds, Ed.), pp. 163-175. San Bernardino County Museum Association Special Publication, San Bernardino, CA.

Kurten, B., and Anderson, E. (1980). "Pleistocene Mammals of North America." Columbia University Press, New York.

Long, A., Hansen, R.M., and Martin, P.S. (1974). Extinction of the Shasta ground sloth. *Geological Society of America Bulletin* **85**, 1843-1848.

Longwell, C. R., Pampeyan, E. H., Bowyer, B., and Roberts, R. J. (1965). Geology and mineral resources of Clark County, Nevada. *Nevada Bureau of Mines Bulletin* **62**.

Martin, P. S. (1973). The discovery of America. *Science* **179**, 969-974.

Martin, P.S. (1984). Prehistoric overkill: the global model. *In* "Quaternary Extinctions" (P. S. Martin and R. G. Klein, Eds.), pp. 354-403. University of Arizona Press, Tucson.

Martin, P.S. (1986). Refuting late Pleistocene extinction models. *In* "Dynamics of Extinction" (D. K. Elliott, Ed.), pp. 107-130. John Wiley & Sons, New York.

Martin, P. S., and Klein, R. G. (Eds.) (1984). "Quaternary Extinctions". University of Arizona Press, Tucson.

Mawby, J. E. (1967). Fossil vertebrates of the Tule Springs site, Nevada. *In* "Pleistocene Studies in Southern Nevada" (H. M. Wormington and D. Ellis, Eds.). *Nevada State Museum Anthropological Papers* **13**, 115-129.

Mayewski, P. A., Meeker, L. D., Whitlow, S., Twickler, M. S., Morrison, M. C., Alley, R. B., Bloomfield, P., Taylor, K. (1993). The atmosphere during the Younger Dryas. *Science* **261**, 195-197.

McVickar, J. L. and Spaulding, W. G. (1993). "Report on the Paleontological and Paleoenvironmental Mitigation of Clark County Regional Flood Control District's North Las Vegas facility." Dames & Moore, Las Vegas.

Meek, N. (1989a). Geomorphic and hydrologic inplications of the rapid incision of Afton Canyon, Mojave Desert, California. *Geology* **17**, 7-10.

Meek, N. (1989b). Physiographic history of the Afton basin, revisited. *In* "The west-central Mojave Desert: Quaternary Studies Between Kramer and Afton Canyon" (R. E. Reynolds, Ed.), pp. 78-83. San Bernadino County Museum Association Special Publication, San Bernardino, CA.

Mehringer, P. J., Jr. (1967). Pollen analysis of the Tule Springs site, Nevada. *In* "Pleistocene studies in southern Nevada" (H. M. Wormington and D. Ellis, Eds.). *Nevada State Museum Anthropological Papers* **13**, 129-200.

Mehringer, P. J., Jr. (1986). Prehistoric environments. *In* "Handbook of North American Indians, Volume 11, Great Basin" (W. L. D'Azevedo, Ed.), pp. 31-50. Smithsonian Institution, Washington, D. C.

Mehringer, P. J., Jr., Martin, P. S., and Haynes, C. V., Jr. (1967). Murray Springs- a mid-postglacial pollen record from southern Arizona. *American Journal of Science* **265**, 786-797.

Mehringer, P. J., Jr., and Ferguson, C. W. (1969). Pluvial occurrence of bristlecone pine (*Pinus aristata*) in a Mohave Desert mountain range. *Journal of the Arizona Academy of Science* **5**, 284-292.

Mehringer, P. J., Jr., and Warren, C. N. (1976). Marsh, dune, and archaeological chronology, Ash Meadows, Amargosa Desert, Nevada. *In* "Holocene Environmental Change in the Great Basin" (R. Elston, Ed.). *Nevada Archaeological Survey Research Papers* **6**, 120-150.

Miller, W. E. (1987). *Mammut americanum*, Utah's first record of the American mastodon. *Journal of Paleontology* **61**, 168-183

Mix, A. C. (1987). The oxygen-isotope record of deglaciation. *In* "North America and Adjacent Oceans During the Last Deglaciation" (W.F. Ruddiman and H.E. Wright, Jr., Eds.), pp. 111-136. The Geology of North America, K-3. Geological Society of America, Boulder, Colorado.

Olausson, E., Ed. (1982). "The Pleistocene/Holocene Boundary in Southwestern Sweden." Sveriges Geologiska Undersoking, Serie C, NR 74, Stockholm.

Ore, H. T., and Warren, C.N. (1971). Late Pleistocene-early Holocene geomorphic history of Lake Mojave, California. *Geological Society of America Bulletin* **82**, 2553-2562.

Paula Cuoto, C. de (1971). On two small Pleistocene ground sloths. *Annals of the Brazilian Academy of Sciences* Suppl. **43**, 499-413.

Petersen, K. L. (1988). Climate and the Dolores River Anasazi. *University of Utah Anthropological Papers* **113**, Salt Lake City.

Porter, S. C., Pierce, K. L. , and Hamilton, T. D. (1983). Late Wisconsin mountain glaciation in the western United States. *In* "The Late Pleistocene" (S. C. Porter, Ed.), pp. 71-113. University of Minnesota Press, Minneapolis.

Quade, J. (1986). Late Quaternary environmental changes in the Upper Las Vegas Valley, Nevada. *Quaternary Research* **26**, 340-357.

Quade, J., and Pratt, W. L. (1989). Late Wisconsin groundwater discharge environments of the southwestern Indian Springs Valley, southern Nevada. *Quaternary Research* **31**, 351-370.

Reynolds, R. E., Reynolds, R. L., Bell, C. J., Czaplewski, N. J., Goodwin, H. T., Mead, J. I., and Roth, B. (1991a). The Kokoweef Cave faunal assemblage. *In* "Crossing the Borders: Quaternary Studies in Eastern California and Southwestern Nevada" (R. E. Reynolds, Ed.), pp. 97-103. San Bernardino County Museum Association Special Publication, San Bernardino, CA.

Reynolds, R. E., Reynolds, R. L., Bell, C. J., and Pitzer, B. (1991b). Vertebrate remains from Antelope Cave, Mescal Range, San Bernadino County, California. *In* "Crossing the Borders: Quaternary Studies in Eastern California and Southwestern Nevada" (R. E. Reynolds, Ed.), pp. 107-109. San Bernardino County Museum Association Special Publication, San Bernardino, CA.

Ritchie, J. C. (1986). Climate change and vegetation response. *Vegetatio* **67**, 65-74.

Rosenzweig, M. L. (1968). Net primary productivity of terrestrial communities: prediction from climatological data. *American Naturalist* **102**, 67-74.

Scuderi, L. A. (1987). Glacier variations in the Sierra Nevada, California, as related to a 1200-year tree-ring chronology. *Quaternary Research* **27**, 220-231.

Shmida, A., Evenari, M., and Noy-Meir, I. (1986). Hot desert ecosystems: an integrated view. *In* "Hot deserts and arid shrublands, B" (M. Evenari *et al.*, Eds.), pp. 379-387. Elsevier, Amersterdam.

Shreve, F. (1942). The desert vegetation of North America. *The Botanical Review* **8**, 195-246.

Simpson, G. G. (1937). A Nevada fauna of Pleistocene type and its probable association with man. *American Museum of Natural History Novitates* **667**.

Smith, G. I. (1979). Subsurface stratigraphy and geochemistry of late Quaternary evaporites, Searles Lake, California. *U.S. Geological Survey Professional Paper* **1043**.

Sorensen, T. (1947). A method of establishing groups of equal amplitude in plant sociology based on similarity of species content. *Det Kongress Danske Videnske, Selskab Biologica Skrifter* **5**,1-34.

Spaulding, W. G. (1984). The last glacial-interglacial climatic cycle: its effects on woodlands and forests in the American West. *In* Proceedings of the Eighth North American Forest Biology Workshop. Department of Forest Resources, Utah State University, Logan.

Spaulding, W.G. (1985). Vegetation and climates of the last 45,000 years in the vicinity of the Nevada Test Site, south-central Nevada. *U.S. Geological Survey Professional Paper* **1329**.

Spaulding, W.G. (1990a). Vegetation dynamics during the last deglaciation, southeastern Great Basin, U. S. A. *Quaternary Research* **33**, 188-203.

Spaulding, W. G. (1990b). Vegetational and climatic development of the Mojave Desert: the last glacial maximum to the present. *In* "Packrat Middens: The Last 40,000 Years of Biotic Change" (J. L. Betancourt, T. R. Van Devender, and P. S. Martin, Eds.), pp. 166-199. University of Arizona Press, Tucson.

Spaulding, W. G. (1990c). Comparison of pollen and macrofossil based reconstructions of late Quaternary vegetation in western North America. *Review of Paleobotany and Palynology* **64**, 359-366.

Spaulding, W. G. (1991). A middle Holocene vegetation record from the Mojave Desert and its paleoclimatic significance. *Quaternary Research* **35**, 427-437.

Spaulding, W. G. (1993). Environmental context. *In* "Kern River Gas Transmission Company, Kern River Pipeline, Cultural Resource Data Recovery Report, California" (R. M. Apple and A. L. York,

Eds.), pp. 7-26. Dames & Moore, San Diego.

Spaulding, W. G., and Martin, P. S. (1979). Ground sloth dung of the Guadalupe Mountains. *In* "Biological investigations in the Guadalupe Mountains National Park, Texas (H. H. Genoways and R. J. Baker, eds.). *National Park Service Proceedings and Transaction Series* **4**, 13-30.

Spaulding, W. G., and Graumlich, L. J. (1986). The last pluvial climatic episodes in the deserts of southwestern North America. *Nature* **320**, 441-444.

Thompson, R. S. (1990). Late Quaternary vegetation and climate in the Great Basin. *In* "Packrat Middens: The Last 40,000 Years of Biotic Change" (J. L. Betancourt, T. R. Van Devender, and P. S. Martin, Eds.), pp. 200-239. University of Arizona Press, Tucson.

Thompson, R.S., Van Devender, T. R., Martin, P. S., Foppe, T., and Long, A. (1980). Shasta ground sloth (*Nothrotheriops shastense* Hoffstetter) at Shelter Cave, New Mexico: environment, diet, and extinction. *Quaternary Research* **14**, 360-376.

Thompson, R. S., and Mead, J. I. (1982). Late Quaternary environments and biogeography in the Great Basin. *Quaternary Research* **17**, 39-55.

Van Devender, T.R. (1987). Holocene vegetation and climate in the Puerto Blanco Mountains, southwestern Arizona. *Quaternary Research* **27**, 51-72.

Van Devender, T.R. (1990). Late Quaternary vegetation and climate of the Sonoran desert, United States and Mexico. *In* "Packrat Middens: The Last 40,000 Years of Biotic Change" (J. L. Betancourt, T.R. Van Devender, and P.S. Martin, Eds.), pp. 134-165. University of Arizona Press, Tucson.

Van Devender, T. R., and Spaulding, W. G. (1979). Development of vegetation and climate in the southwestern United States. *Science* **204**, 701-710.

Van Devender, T.R., Spaulding, W.G., and Phillips, A. M., III (1979). Late Pleistocene plant communities in the Guadalupe Mountains, Culberson County, Texas. *In* "Biological Investigations in the Guadalupe Mountains National Park, Texas" (H. H. Genoways and R. J. Baker, Eds.). *National Park Service Proceedings and Transaction Series* **4**, 13-30.

Van Devender, T. R., Thompson, R. S., and Betancourt, J. L.(1987). Vegetation history of the deserts of southwestern North America: the nature and timing of the Late Wisconsin-Holocene transition. *In* "North America and Adjacent Oceans During the Last Deglaciation" (W. F. Ruddiman and H. E. Wright, Jr., Eds.), pp. 323-352. The Geology of North America, vol. K-3. Geological Society of America, Boulder, Colorado.

Webb, R.H., and Betancourt, J. L. (1990). The spatial and temporal distribution of radiocarbon ages from packrat middens. *In* "Packrat Middens: The Last 40,000 Years of Biotic Change" (J. L. Betancourt, T. R. Van Devender, and P. S. Martin, Eds.). pp. 85-102. University of Arizona Press, Tucson.

Wells, P.V., and Woodcock, D. (1985). Full-glacial vegetation of Death Valley, California- Juniper woodland opening to *Yucca* semidesert. *Madrono* **32**, 11-23.

Wells, S. G., McFadden, L. D. , and Dohrenwend, J. C. (1987). Influence of late Quaternary climatic changes on geomorphic and pedogenic processes on a desert piedmont, eastern Mojave Desert, California. *Quaternary Research* **27**, 130-146.

Wells, S. G., Anderson, R. Y. , McFadden, L. D. , Brown, W. J., Enzel, Y. and Miossec, J-L. (1989). Late Quaternary paleohydrology of the eastern Mojave River drainage, southern California: quantitative assessment of the late Quaternary hydrologic cycle in large arid watersheds. *New Mexico Water Resources Research Institute Report* **242**.

Whittaker, R. H. (1967). Gradient analysis of vegetation. *Biological Review* **42**, 207-264.

Winograd, I. J., and Thordarson, W. (1975). Hydrogeologic and hydrochemical framework, south-central Great Basin, with special reference to the Nevada Test Site. *U.S. Geological Survey Professional Paper* **712-C**.

Woodcock, D. (1986). The late Pleistocene of Death Valley: a climatic reconstruction based on macrofossil data. *Palaeogeography, Palaeoclimatology, Palaeoecology* **57**, 273-283.

Wright, H. E., Jr. (1984). Sensitivity and response time of natural systems to climatic change in the Late Quaternary. *Quaternary Science Reviews* **3**, 91-132.

York, A. L. (1993). " Class III cultural resources inventory for Los Angeles Department of Water and Power Mead to Adelanto Transmission Line Project: Cronese and Calico Divisions." Dames & Moore, San Diego.

Chapter 12

Pika (*Ochotona*) and Paleoecological Reconstructions of the Intermountain West, Nevada and Utah

JIM I. MEAD* and W. GEOFFREY SPAULDING†

**Quaternary Studies Program and Department of Geology, Northern Arizona University, Flagstaff, Arizona 86011-4099; The Mammoth Site, Hot Springs, South Dakota 57747*
†Dames & Moore, 4220 S. Maryland Parkway, Suite 108, Las Vegas, Nevada 89109

INTRODUCTION

We review evidence for the late Quaternary distribution and paleoecology of pika (Lagomorpha: Ochotonidae: *Ochotona princeps*) in the Intermountain West of Utah and Nevada, western North America (Fig. 1). Dung pellets of pika have been recovered from radiocarbon dated packrat (Rodentia: Muridae, *Neotoma* spp.) den accumulations (middens) of middle and late Pleistocene age (late Wisconsinan Glacial age). Analysis of these records was undertaken to produce a: 1) chronologically controlled paleo-zoogeographic study, 2) dietary reconstructions of the pika, 3) characterization of the local paleoecology of glacial-age pika habitats, and 4) comparison of local plant community utilization by two commensals, *Neotoma* and *Ochotona*.

The pika is a small lagomorph with many species distributed throughout the Holarctic region. Unfortunately, there are few usable criteria for adequately recognizing subgenera of *Ochotona* (Smith *et al.*, 1990). Because of the considerable geographical and seasonal differences within forms of *Ochotona*, there is wide debate on the actual number of living species (Corbet, 1978; Mitchell, 1981; Smith *et al.*, 1990). Only two forms, *O. princeps* (American pika) and *O. collaris* (collared pika) live in North America (Hall, 1981).

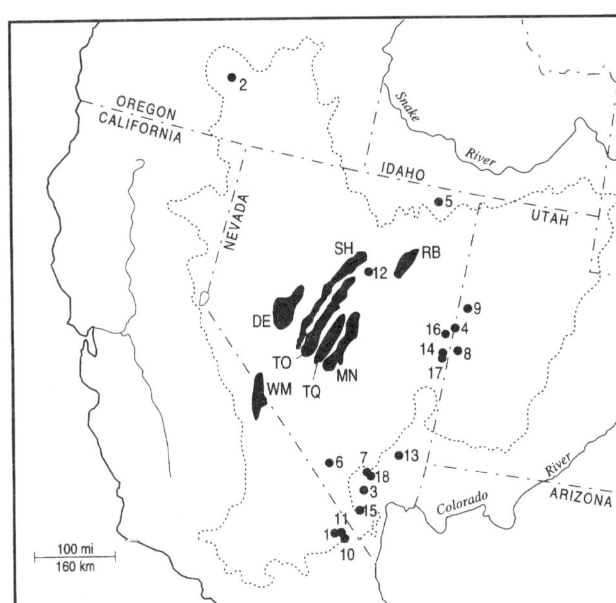

FIG. 1. The Intermountain West, locating the modern distribution and late Pleistocene occurrences of *Ochotona princeps*.

Late Quaternary Environments and Deep History: A Tribute to Paul S. Martin, edited by David W. Steadman and Jim I. Mead. The Mammoth Site of Hot Springs, South Dakota, Inc. Scientific Papers, Volume 3. Hot Springs, South Dakota, 1995.

The American pika is considered to be part of the Alpina complex which also includes the two Asian forms, *Ochotona alpina* (alpine or Altai pika) and *O. hyperborea* (northern or Siberian pika), and the North American species *O. collaris*. The two American forms are distinct from the Asian forms in morphology (Weston, 1981) and chromosomal arrangement (Smith *et al.*, 1990).

The species living today in the Intermountain West and adjacent areas is *O. princeps*. There is no indication from the late Pleistocene record that any other species has also lived in this region. Mead (1987) provided a review of Quaternary *Ochotona* localities of North America, which includes the Intermoutain West. The only new pika localities in the Intermountain West since that list was published (and reused by Hafner, 1993) are from Snake Creek Burial Cave, Snake Valley (Mead and Mead, 1989), Cathedral Cave, Snake Range (Bell, 1990; Mead *et al.*, 1992), and Forty Mile Canyon, North Yucca Mountain (this report), all localities in Nevada. All living American species are considered to be small in size, compared to some extinct species that were much larger, both in North America and Asia (Churcher and Dods, 1979; Erbajeva, 1988; Guthrie and Matthews, 1971; Smith *et al.*, 1990). Large forms of *Ochotona* have been extinct in North America since the Wisconsinan. (Mead and Grady, ms).

Habitats

North American pika presently live in regions among predominantly boreal and alpine vegetation (Orr, 1977). Smith *et al.* (1990:45) indicate that the species is "specific to talus or piles of broken rock." It seems that, presently, the American pika is restricted to rocky habitats similar to those shown in Figures 2 and 3. Their occasional excursions today to lower elevations seem to be restricted by the presence

FIG. 2. Example of a modern talus slope community inhabited by *Ochotona princeps*, Cedar Breaks, Utah.

FIG. 3. *Ochotona princeps* habitat at Craters-of-the-Moon, Idaho.

of high summer temperatures (Smith, 1974). In the west, pika live in the mountainous areas of the Sierra Nevada mainland, Wasatch mainland (including the Uinta Mountains and the Paunsaugunt and Aquarius plateaus), and in the boreal island mountains (in terms of montane island biogeography; Brown, 1978) of the White-Inyo, Desatoya, Toiyabe-Shoshone, Toquima-Monitor, and Ruby ranges (Hall, 1946).

Mead (1987:169) proposed that *Ochotona* was not restricted to, and need not be an indicator of, "a rocky/talus" habitat during the Pleistocene, and therefore, the species is not necessarily an indicator of that ecological habitat. Hafner (1993:379) strongly disagreed and provided an alternative proposal that Pleistocene "remains of *O. princeps* are strong indicators of a cool, mesic, *rocky* habitat". Some of this disagreement comes from the confusion surrounding the interpretation of "a rocky/talus habitat." We wish to clarify Mead's (1987) definition by placing emphasis on "talus", and deleting the word "rocky". Today the American pika is restricted to areas where boulders and rocks are large enough to allow the pika to maneuver in and around the boulder piles. Figure 2 illustrates a typical alpine talus slope that presently supports a population of pika. American pika can also be found in lava flow boulder networks, illustrated here by those at the Craters of the Moon, Idaho (Fig. 3).

The problem in incorporating the word "rocky" in the description of pika habitat requirements is that it is too general a term and leads to confusion. The great majority of terrains in western North America are "rocky". The critical characteristic of modern pika habitat appears to be the need to maneuver in, around, and *through* rock cover (Fig. 2). A habitat consisting of a limited pile of rocks, or a rock shelter, that does not permit the maneuverability mentioned above, and is surrounded by tens of kilometers of alluvial fan, does not seem to be part of the habitat requirements of the American pika today.

Hafner (1993) apparently interpreted the word 'rocky', as used by various authors, to indicate a maneuverable/livable bouldery talus pile. Mead (1987) cited the occurrence of Pleistocene pika remains found in valley-bottom situations. Although rocky, some localities do not contain a maneuverable/livable talus slope in the area (Fig. 4). The same situation exists for northern Mojave Desert fossil localities such as the Eleana Range and Penthouse sites (Table 1, Fig. 5; also see Spaulding, 1985: fig. 15). These localities containing fossils of pika are apparently not suitable for *Ochotona* under today's conditions.

Dung pellets

Hafner (1993:377) argues that "small (4.5 mm diameter)...round, and buoyant pika dung pellets could easily have been washed [down] from steep rocky slopes. A talus tends to be a high-energy environment that partly depends on erosion and hydraulic flushing to prevent soil buildup. Floating pellets may have been further sorted and concentrated due to their uniform size and density and selectively incorporated into *Neotoma* middens." Here he appears to be trying to make a case that all late Pleistocene records of *Ochotona* must be indicative of talus habitat.

The dung pellet records reported by Mead (1987) and herein are contrary to Hafner's (1993) proposal that the dung must have washed down out of the "rocky" mountains to settle in non-rocky, atypical habitat areas. Pellets transported in such a high-energy environment are often torn apart or abraded, and ultimately sun-bleached. Their mucosal coating is removed by such exposure. Water that transports dung would initiate microbial decay by soaking the remains. Bile acids (primary and secondary) are easily leached from dung, therefore removing any nuclear magnetic resonance (NMR) signature obtainable from the coproli-

FIG. 4. The Garrison locality, Snake Valley, Nevada/Utah, illustrating the lack of talus slopes (as described in text) that *Ochotona* seems to require today.

TABLE 1. Radiocarbon dated *Neotoma* middens in the Intermountain West, North America, containing dung pellets of *Ochotona*. All radiocarbon dates are on plant remains from packrat middens, unless otherwise noted. D, dung pellet. S, skeletal remains. !, not as published in Spaulding (1981); apparently mixed, glacial-age assemblage. *, dung pellets of *Ochotona* recovered from the *Neotoma* midden, although it is not known which ^{14}C date is best associated with the dung.

Locality	^{14}C years B.P. (lab #)	Pika Specimen	Reference
	Northern Mojave Desert		
SHEEP RANGE			
Spires 2	22,200±800 (AA-689)	D*	1
	30,700±800 (AA-688) !		
Flaherty Mesa 1	20,390±340 (WSU-1862)	D & S	1
Penthouse 1	19,400±300 (A-1772)	D	1
Willow Wash 4E	>44,600 (USGS-398)	D	1
SPRING RANGE			
Potosi Mountain 2	14,270±220	S	2
	14,900±180	S	2
Corn Creek 3	14,300±800 (AA-782) (*Ochotona* dung)	D	3
ELEANA RANGE			
ER 2-7	14,760±100 (USGS-887)	D*	4
	15,380±100 (USGS-888)		
ER 2-10	17,100±130 (USGS-1069)	D	4
ER 2-11	10,800±90 (USGS-892)	D*	4
	14,310±110 (USGS-893)		
ER 3-10	32,500±800 (USGS-993)	D*	4
	35,450±650 (USGS-1265)		
	47,100+2600-2000 (USGS-919)		
YUCCA MOUNTAIN			
Fortymile Can. 11	16,410±70 (Ql-4234)	D	this report
	Central Great Basin		
DEEP CREEK RANGE			
Granite Canyon 1	13,620±700 (A-2432)	D	5
SNAKE RANGE			
Arch Cave 2A	34,040+5400-3200(A-2782)	D	5
Smith Creek 4	12,235±395 (GX-5863)	D	6
Smith Creek 5	13,340±430 (A-2094)	D	6
Stream View 1	11,010±400 (A-2095)	D	6
Stream View 2	17,350±435 (GX-5866)	D	6
Stream View 3	8030±510 (AA-432) (*Ochotona* dung)	D	6
SNAKE VALLEY			
Garrison 1	12,230±180 (A-2312)	D & S	7
Garrison 2	13,480±250 (A-2313)	D	7

References: 1, Spaulding (1981); 2, Mead and Murray (1991); 3, Van Devender *et al.* (1985); 4, Spaulding (1985); 5, Thompson (1984); 6, Thompson (1979); and 7, Thompson and Mead (1982).

FIG. 5. The Eleana Range, northern Mojave Desert, Nevada, illustrating the lack of talus slopes (as described in text) that *Ochotona* seems to require today.

tes (J. Theis, personal communication). Pleistocene pika pellets from packrat middens yield distinct NMR bile acid signatures, indicating that they have not been washed by water (to be discussed later). Dung pellets of *Ochotona* recovered from the packrat middens show little degradation. We see no reasonable alternative to our postulate that the late Pleistocene dung pellets originate locally.

TAPHONOMY AND PACKRAT MIDDENS

Packrat middens have proven instrumental in reconstructing late Quaternary plant community changes in the arid Southwest and Intermountain West (Betancourt *et al.*, 1990; Spaulding *et al.*, 1983; Thompson, 1984; Van Devender and Spaulding, 1979; Van Devender *et al.*, 1987). All *Neotoma* are ardent collectors of plants growing less than 100 m from their den (Bleich and Schwartz, 1975). These remains are brought back to the home site for food and nest-covering construction material. This procuring and browsing by packrats does not consistently produce midden assemblages that quantitatively correspond to the actual representation of species in the plant community (Spaulding *et al.*, 1990). Packrat middens do provide a reconstruction of the mosaic of local plant communities, although they are often biased toward browse (woody) species. Pika, in contrast, consume both graze and browse, often sampling species not necessarily collected by the packrat.

Both plant and animal remains are recovered in modern and ancient packrat middens (see various articles in Betancourt *et al.*, 1990). Packrats collect animal remains for den construction, oddity factor, and for scent obscuring from predators. Packrats can procure raptor pellets (stomach contents), isolated carcasses, and dung from various animals. The large bushy-tailed packrat (*Neotoma cinerea*) can collect skeletal remains as large as immature skulls of the extinct mountain goat (*Oreamnos harringtoni*) and the radius-ulna of an adult extinct small (burro-sized) horse (*Equus* sp.) (JIM, personal observations). Although skeletal remains recovered from packrat middens may be local or extralimital (e.g., via raptor pellets), dung (especially of small herbivores) is recovered from the adjacent, local community(ies).

DUNG ANALYSES

Dung from middens

Packrat middens used in paleo-environmental reconstructions of the Mojave Desert and the Great Basin (Spaulding, 1990; Thompson, 1990) have also yielded dung pellets of lagomorphs, including pika (Fig. 1). It is these pika remains that we use in our analyses. Dung pellets of modern and fossil pika and various modern rabbits (Leporidae) were measured for size to facilitate the identification of fossil material. Representative modern and fossil dung pellets were analyzed for bile chemistry, using nuclear magnetic resonance (NMR; by J. N. Theis, University of California, Davis) to begin to develop bile acid signatures for identification of the producer (Theis *et al.*, 1988).

When the abundance of fossil dung pellets permitted, samples were analyzed for microhistologcial contents (by Theresea Foppe, AFAB, Ft. Collins, CO), to be used in dietary reconstructions of modern and fossil pika. Age of the pika remains is assumed to be the same as that of the packrat midden containing the dung pellets, which is based on the radiocarbon dating of either *Neotoma* dung pellets or of a single plant species (Table 1) or, in rare cases, the direct radiocarbon dating of pika dung. Paleoenvironmental reconstructions are made using the packrat midden macrobotanical data. Here we compare the plants harvested by the pika with those procured by the packrat.

Dung morphology and identification

Dung of lagomorphs is different in shape from that typical of rodents such as the packrat. Most dung pellets produced by rodents are tubular, with the length often more than twice as long as the width. Lagomorphs produce an amorphous dung mass that is reinjested to process additional nutrients. Ultimately a dung pellet is produced that is disc shaped in plan view, with length actually being the shortest dimension, and the two diameters of approximate equal proportions. The following discussion is concerned with only the diameter dimensions.

Dung pellets of living *Ochotona* are typically smaller than dung produced by the leporids *Sylvilagus* (cottontails) and *Lepus* (jackrabbits). Dung pellets produced by *Lepus* are much larger than any produced by *Ochotona,* and will not be considered here. While most dung produced by *Sylvilagus* is consistently larger than in *Ochotona*, there is some overlap in size with the smallest leporid, *Brachylagus idahoensis* (pygmy rabbit; sometimes included in the genus *Sylvilagus*).

Dung pellets of living *Ochotona princeps, Brachylagus idahoensis*, and *Sylvilagus nuttalli* were gathered from various localities in western North America for comparisons of size (Table 2). Besides averaging larger than in *B. idahoensis*, dung pellets of *S. nuttalli* have angular facets on the surface that give their irregular-shaped fecal discs a diagnostic shape.

Although most dung pellets of *Brachylagus idahoensis* are larger (>4.8 mm) than those of *Ochotona* (ranging from 2.0 to 4.8 mm), the largest pellets of *Ochotona* and the smallest pellets of *Brachylagus idahoensis* overlap between 4.1 and 4.8 mm (Table 2; Fig. 6.). Dung pellets smaller than 4.1 mm in greatest diameter consistently belong to pika and, therefore, that is our major criterion used in the identification of the fossil dung pellets. Most of the fossil lagomorph dung pellets recovered from the packrat middens are less than 4.0 mm, and clearly belong to *Ochotona* (Fig. 6). Over 100 dung pellets were recovered from the Garrison 1 packrat midden; none measured larger than 4.0 mm. Thirty of the largest dung pellets in Streamview 2 measured between 3.1 and 4.4 mm (mean = 3.8). Most of these seem correctly identified as *Ochotona*, although some conceivably could belong to juveniles of *Brachylagus idahoensis* (Fig. 6).

To provide an independent assessment of the identifications based on the morphology, we submitted modern and fossil dung samples for analysis of the bile acids using NMR. The use of this technique is restricted to well preserved dung pellets, both modern and fossil. Bile acids are easily leached from the pellets by water. The fact that the fossil dung pellets contained bile acids usable in the NMR tech-

TABLE 2. Measurements of dung pellets of *Sylvilagus*, *Brachylagus*, and *Ochotona* (fossil and modern). The smallest measurement is the length of the pellet and therefore is not used here. The two diameters of the disc-shaped dung pellets are near equal in size. Only the larger diameter will be presented. Measurements in mm. "Samples" refers to the number of samples of dung taken from a given locality.

Species	Locality	N	OR	X	SD	Samples
MODERN						
Sylvilagus nuttalli						
	Big Horn Mts., WY	15	6.2-7.7	7.5	0.4	2
	Snake Range, NV	30	6.2-9.7	8.0	0.9	1
Brachylagus idahoensis						
	Bonneville Co., ID	10	5.0-5.8	5.4	0.3	1
	Research Lab, ID	83	4.1-9.0	5.9	0.9	?
	Schell Creek Mts., NV	30	4.4-6.5	5.3	0.6	1
Ochotona princeps						
	Beartooth Mts., WY	10	2.9-4.1	3.4	0.4	1
	Cedar Breaks, UT	10	3.0-4.6	3.8	0.6	3
	Craters-of-the-Moon, ID	10	2.1-4.0	3.2	0.6	4
	Mt. Rainier, WA	40	2.6-4.8	4.2	0.4	1
	Ruby Mts., NV	15	2.8-3.9	3.5	0.3	5
	Sequoia, CA	20	3.5-4.5	4.1	0.4	1
	Virginia Basin, CO	10	2.6-3.5	3.1	0.3	1
	White Mts., CA	20	3.5-4.5	3.9	0.3	2
PLEISTOCENE						
Ochotona cf. *O. princeps*						
	Arch 2A	1	3.0	-	-	-
	ER 2-7	1	3.8	-	-	-
	ER 2-10	11	3.0-3.9	3.4	0.3	-
	ER 2-11	40	2.8-4.4	3.6	0.4	-
	ER 3-10	1	3.7	-	-	-
	Fortymile 11A(2)	2	3.2	3.2	0	-
	Garrison 1	12	3.0-4.0	3.5	0.5	-
	Garrison 2	10	2.0-3.7	3.3	0.5	-
	Granite 1	8	2.2-3.8	3.2	0.5	-
	Smith Creek 4	16	3.0-4.5	3.5	0.4	-
	Smith Creek 5	2	3.0-3.5	3.2	0.3	-
	Spires 2(2)	15	2.9-4.0	3.4	0.4	-
	Streamview 1	1	3.6	-	-	-
	Streamview 2	30	3.1-4.4	3.8	0.3	-
	Streamview 3	3	3.7-4.1	3.9	0.2	-
	Willow Wash 4E2	10	2.8-3.6	3.1	0.3	-

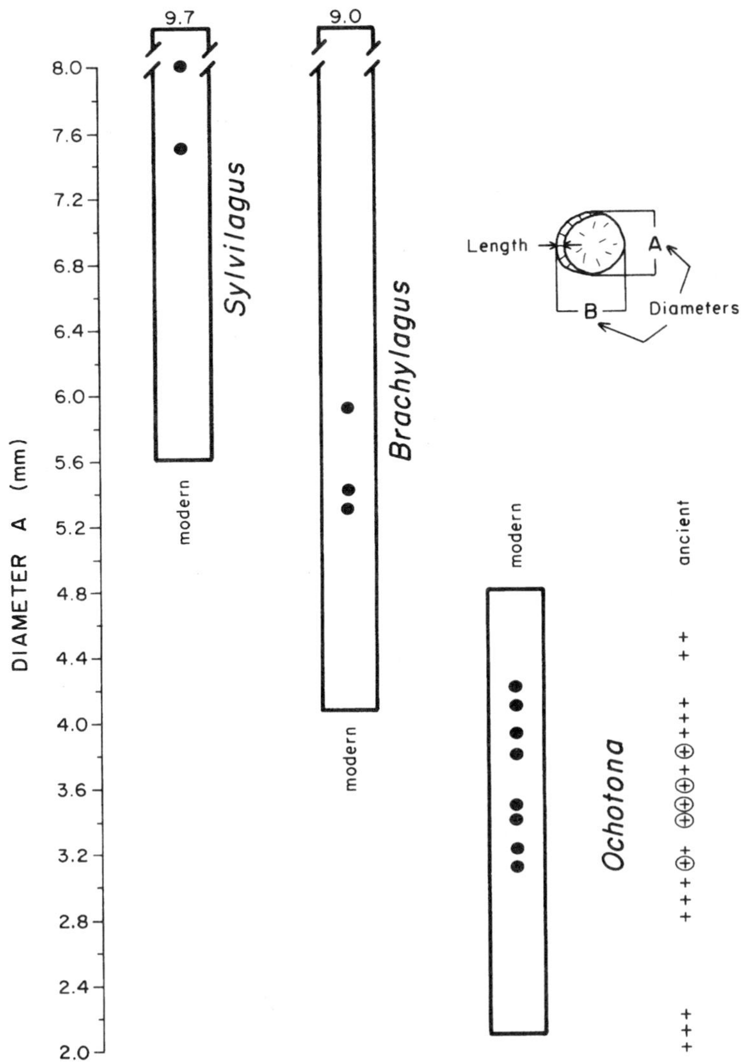

FIG. 6. Measurements of dung pellets in *Ochotona*, *Brachylagus*, and *Sylvilagus*.

nique attest to their good preservation. This technique of obtaining bile chemical NMR "signatures" for dung (especially with fossil dung) is in its infancy. Figure 7 illustrates the NMR profiles of selected modern and fossil lagomorph dung pellets. Although preliminary, it appears to indicate that chemical spikes at 1.05 and 1.35 may be indicative of *Ochotona*. These spikes appear with modern pika samples and the single assayed fossil sample from Streamview 2, and not with *Sylvilagus*. This preliminary effort at using NMR to identify fossil dung would seem to verify that the Streamview 2 locality does contain *Ochotona* dung pellets.

THE PALEOENVIRONMENT OF INTERMOUNTAIN PIKA

Table 1 lists the packrat middens that contain *Ochotona* fossils, both skeletal and dung remains. Because the skeletal specimens of pika could have been introduced into the locality via a raptor pellet or carnivore scat, and therefore represent a possible extralimital species, we feel that it is best to limit the discussion of paleocommunites that contained pika to those packrat middens that yielded dung pellets (which omits the Potosi Mountain middens). For reasons articulated below, it is our judgement that the recovery of *Ochotona* dung pellets in the packrat middens is an indication that the animal actually lived in the local community. Pika have not been recovered from packrat middens where the genus presently lives today; all localities to be discussed represent extralimital records. Pika dung pellets have been recovered from packrat middens in two major regions, the northern Mojave Desert and the east-central Great Basin Desert. Discussions will be divided into these two different environmental and physiographic regions.

As we will illustrate, plant macrofossil assemblages from packrat middens reflect primarily the species that formed the shrub and tree strata of the site's plant community, and it is these strata that are "seen" least by the small, den-dwelling pika. Nevertheless, the midden assemblages yield valuable data on the local environment (within 100 m of the site and, primarily within 30 m of the site [Spaulding *et al.*, 1990]). When combined with dietary analyses on pika dung pellets, they provide a more comprehensive understanding of pika habitat during the last glacial.

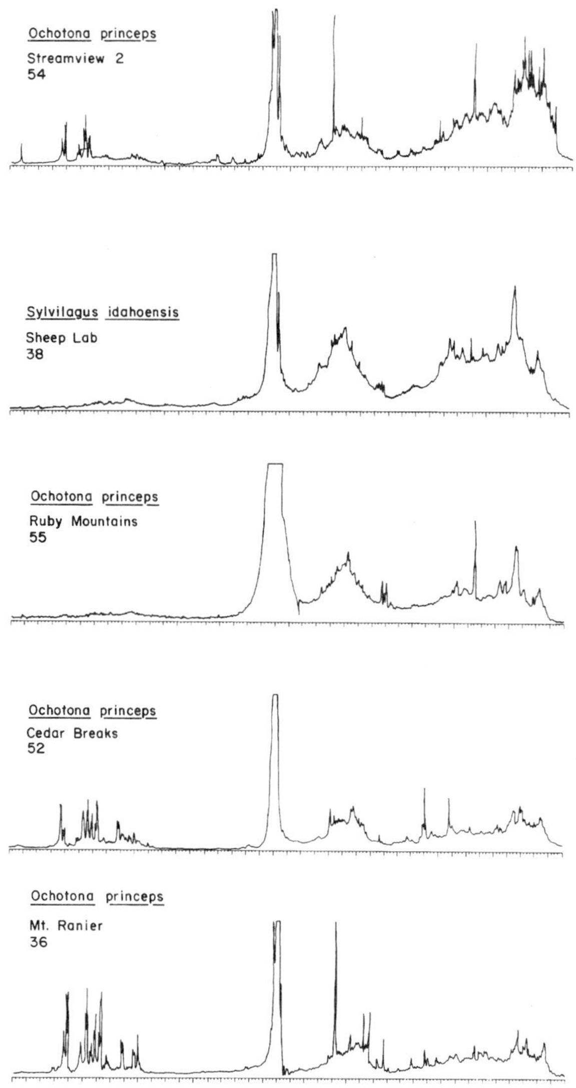

FIG 7. Nuclear magnetic resonance profiles (signatures) of bile acids recovered from modern and fossil dung pellets of *Ochotona*, and *Sylvilagus*.

Northern Mojave Desert

Packrat middens yielding pika dung pellets were recovered from a variety of sites in this region. Most are part of the large mountain masses of the Spring and Sheep ranges (Table 1). The comparatively low Corn Creek 3 site is an inselberg of the Spring Range, and is separated from the main mountain mass by about 10 km of alluvial fan. The Eleana Range and Fortymile Canyon lie in the complex highland area south of Pahute Mesa and the Belted Range, the latter extending north into the Great Basin Desert (Beatly, 1976; Fig. 1). With the possible exception of Potosi Mountain 2, all these sites are near the tops of ridges or mesas, and they fail to meet the topographic criteria necessary to promote the downward movement of pika dung pellets from much higher elevations, proposed by Hafner (1993) as an alternative explanation to account for their presence at low elevations (Fig. 5).

These Mojave Desert sites cover a broad elevational range, from 1160 to 2040 m. The lower latitude of these sites, and the lower elevations of the adjacent valleys relative to those of the Great Basin Desert farther north, mean that a wider array of vegetation types occurred here during the last glacial age. The highest elevation midden sites from the northern Mojave Desert yield macrofossil assemblages indicating a subalpine woodland with a diverse shrub component. Below ca. 1800 m elevation on the mountain slopes, subalpine conifers gave way to Utah juniper (*Juniperus osteosperma*). At even lower elevations, a steppe shrub-woodland mosaic existed on rocky slopes, and steppe, or juniper-sagebrush savanna, may have been continuous across the vast bajadas (Mehringer, 1967; Spaulding, 1990; Thompson and Mead, 1982).

What is intriguing about the pika records from the northern Mojave Desert is that they indicate, beyond any doubt, that the species extended its range well into the lower, relatively xeric vegetation zones. Based on these data, it thrived more than 600 m below the glacial-age lower limit of subalpine woodland and, therefore, the habitat breadth of pika was broader during the last glacial age than it is today. *Ochotona* is presently restricted to alpine and subalpine vegetation zones, and we know of no areas where it can be found today living in habitats characterized by Utah juniper and sagebrush (*Artemisia* subgen. *Tridentatae*; Table 3), as it did during the glacial age.

Central Great Basin

During the Wisconsinan Glacial age, an ice cap was present at higher elevations in the Snake Range. During this same period, the lowlands of the Snake Valley were inundated by Lake Bonneville, resulting in the compression of terrestrial communities between the extremes of glacier and lake. The pika-dung bearing middens from this region document a minimum elevational range for the species of only 430 m, in contrast with a minimum range of 880 m for the northern Mojave Desert (Tables 3, 4). This is due to the restriction posed by the much higher elevations of these northern valleys. For example, the bottom of the Upper Las Vegas Valley in the vicinity of the Corn Creek 3 site is ca. 1030 m elevation, while the bottom of Snake Valley in the vicinity of the Garrison locality is about 1600 m eleva-

TABLE 3. Summary of floristic data from *Neotoma* midden samples yielding *Ochotona* dung pellets recovered from the northern Mojave Desert and vicinity. See Table 1 for references and radiocarbon dates.

Site	Elev. (m)	Woodland type	Abundant Species	Common Species
		SHEEP RANGE		
Spires 2	2040	subalpine conifer	Pilo, Pifl	Fone, Juos, Cein
Flaherty Mesa 1	1770	juniper	Juos	-
Penthouse 1	1600	juniper	Juos	Artr, Fapa, Fone
Willow Wash 4	1585	juniper	Juos	Chvi, Pimo, Sym, Teca
		SPRING RANGE		
Potosi Mt. 2	1880	subalpine conifer	Pilo, Pifl	not reported
Corn Creek 3	1160	juniper	Juos	not reported
		OTHER LOCALITIES		
Eleana Range 2	1810	subalpine conifer	Pifl	Artr, Lu, Poa, Teca
Eleana Range 3	1800	subalpine conifer	Pifl	Artr, Sym, Cele, Chvi
Fortymile Can. 11	1310	juniper-sagebrush	Artr	Cele, Juos, Oppo

Abbreviations: Artr, *Artemisia* subgen. *Tridentatae*; Cein, *Cercocarpus intricatus*; Cele, *C. ledifolius*; Chvi, *Chrysothamnus viscidiflorus*; Fapa, *Fallugia paradoxa*; Fone, *Forsellesia nevadensis*; Juos, *Juniperus osteosperma*; Lu, *Lupinus* cf. *argenteus*; Oppo, *Opuntia polyacantha*-type; Pifl, *Pinus flexilis;* Pilo, *P. longaeva*; Pimo, *P. monophylla;* Poa, Poaceae; Sym, *Symphoricarpos* sp.; Teca, *Tetradymia canescens*-type.

tion. No doubt the potential habitat range of pika during the last glacial age extended far below the lowest elevation midden site in the central Great Basin.

The topography of these midden sites can be more extreme than those of the northern Mojave Desert. However, the maximum vertical distance that pika dung pellets could have been washed down slope to the middens sites, with Hafner's (1993) concerns in mind, is <100 m in the case of the Streamview, Smith Creek Cave, and the Garrison sites. This latter locality is a true inselberg, and is separated from the Snake Range to the west by ca. 5 km of alluvial fan. The nearby Burbank Hills afford no suitable modern habitat for the pika, in relation to Hafner's (1993) concerns regarding the possibility of long distance transport of pellets. Although the substrate was rocky, this apparently was not an area of talus slopes and rock piles typical of modern pika habitat (Figs. 4 and 8).

TABLE 4. Summary of floristic data from *Neotoma* middens assemblages that yield *Ochotona* dung pellets recovered from the central Great Basin. See Table 1 for references and radiocarbon dates.

Site	Elev. (m)	Woodland type	Abundant Species	Common Species
		DEEP CREEK RANGE		
Granite Can. 1	2070	subalpine conifer	Pifl, Juco	Artr
		SNAKE RANGE		
Arch Cave 2A	1980	mixed conifer	Juos, Pilo, Artr, Sym	Cact, Chry
Smith Creek 4	1950	subalpine conifer	Pilo, Artr,Sym	Chry
Smith Creek 5	1950	subalpine conifer	Pilo, Artr	Chry
Streamview 1	1860	subalpine conifer	Pilo, Artr, Chry	-
Streamview 2	1860	subalpine conifer	Pien, Pilo	Juco, Artr, Chry, Fone, Rimo, Sym, Cirs, Phlo, Viam
Streamview 3	1860	mixed conifer	Juos, Artr, Epvi, Fone	Pilo, Cact, Cela, Or
		SNAKE VALLEY		
Garrison 1	1640	subalpine conifer	Pifl, Artr	Chen, Chry, Sym
Garrison 2	1640	subalpine conifer	Pifl, Artr	Chen, Sym

Abbreviations: Artr, *Artemisia* subgen. *Tridentatae*; Atco, *Atriplex confertifolia*; Cact, Cactaceae; Cein, *Cercocarpus intricatus*; Cela, *Ceratoides lanata*; Chen, *Chenopodium* sp.; Chry, *Chrysothamnus* sp.; Cirs, *Cirsium* sp.; Epvi, *Ephedra* cf. *viridis*; Fone, *Forsellesia nevadensis*; Gusa, *Gutierrezia* cf. *sarothrae*; Juco, *Juniperus communis*; Juos, *J. osteosperma*; Or, *Oryzopsis hymenoides*; Phlo, *Phlox* sp.; Pien, *Picea engelmannii*; Pifl, *Pinus flexilis;* Pilo, *P. longaeva*; Rimo, *Ribes montigenum*; Sym, *Symphoricarpos* sp.; Viam, *Vicia* cf. *americana*.

FIG. 8. Habitat around and between the Garrison and Snake Creek Burial Cave localities, just above the high level of glacial Lake Bonneville. Late Pleistocene *Ochotona* lived in this region in a xeric subalpine woodland with shrubs.

Detailed descriptions of the changing environments of the Great Basin Desert for the past ca. 35,000 yr are provided by Thompson (1990) and reviewed by Grayson (1993). Most central Great Basin middens yielding pika dung also contain plant macrofossils indicating subalpine conifer woodland typified by bristlecone pine (*Pinus longaeva*; Table 4). Other conifers in this vegetation zone included limber pine and Engelmann spruce (*Picea engelmannii*). A diverse array of shrub species is also represented in the middens, particularly sagebrush, rabbitbrush (*Chrysothamnus* sp.), and snowberry (*Symphoricarpos* sp.). Only two midden assemblages suggest vegetation other than subalpine conifer woodland, those from Arch Cave 2A and Streamview 3 (Table 4). Interestingly, these are also the only two middens that are not late Wisconsinan in age. The former dates to the middle Wisconsinan interstadial at approximately 34,000 yr B.P., while the latter dates to the last millennium of the early Holocene (ca. 8000 yr B.P.; Table 1). They yield macrofossils indicating open, mixed conifer or juniper woodland (Table 4). Indeed, the relative abundance of heliophytic taxa in the assemblages (*e.g.*, sagebrush, rabbitbrush), demonstrate that open habitat prevailed in the vicinity of sites located within subalpine conifer woodland, as well as sites within mixed conifer or juniper woodland.

Smith Creek Canyon provides a unique situation for the region. The canyon contained a glacial meltwater stream, providing a local environment that permitted colonization not only by subalpine conifers and steppe shrubs, but also by phreatophytic species. The actual stream-side area supported riparian species including water birch (*Betula occidentalis*) and willow (*Salix* sp.) (Thompson, 1979). This mesic canyon-bottom habitat, as well as the north-facing hillside, was most likely the restricted location for the northern bog lemming (*Mictomys borealis*) that was recovered from Cathedral and Smith Creek caves. But it also provided one of the few mesic localities in the region for the pika (Mead *et al.*, 1992). Based on the present environmental requirements of the pika, one would expect that *Ochotona* lived in this canyon during the glacial, as indeed they did.

DIETARY RECONSTRUCTION OF THE PLEISTOCENE PIKA

While the habitat breadth of the pika expanded radically, relative to the present, during the last glacial age, dietary analyses suggest that its food habits may have remained largely the same as those populations inhabiting high elevations in the central Great Basin today. We find no evidence for a broadening or shift in the pika food niche that would be at all comparable to the apparent change in its habitat breadth. As now, the pika apparently relied on a predictable harvest of forbs and grasses to harvest during the cool summer months of the last glacial age (Huntly, 1987; Huntly *et al.*, 1986; Millar and Zwickel, 1972; Smith and Weston, 1990).

The current diet of the pika in the Intermountain West and adjacent areas today was characterized by the microhistological analysis of contents in dung pellets collected from the Ruby Mts. (NV), White Mts. (CA), Craters-of-the-Moon (ID), Cedar Breaks (UT), and Bear Tooth Mts. (WY) (Table 5). Only in the Ruby Mountains did graminoids dominate (57.8%) over the browse (42.2%)(Fig. 9). In all cases, graze is important in the modern diet of the pika (57.8% to 24.3% in 5 samples). Table 6 lists the pika samples analyzed for microhistological remains recovered from late Pleistocene packrat middens (SCC 4, GN 2, GC 1, Stv 2, FMC 11A(2), Sp 2(2), and WW 4E2), ranging in age from 12,230 to >44,600 yr B.P. Late Pleistocene samples of pika dung contained 0% to 27.6% graze species (Fig. 9).

Pika, like packrats, are generalized herbivores. Differences in composition of plants harvested are frequent and reflect local variation in the structure of the nearby plant community (Orr, 1977; Smith *et al.*, 1990). However, in contrast to the collecting of the packrat, pika appear to harvest a wider array of grasses.

In all samples a greater number of plant taxa were identified from the packrat midden (macrobotanical remains) than from the pika dung pellet(s) (microhistological remains). The total number of encountered plant species is provided for both pika and packrats from the same fossil unit (Table 7). It is apparent that macrofossils from the packrat midden provide a more detailed account of the local plant community at any given time than does the analysis of the pika dung (Table 6). Table 7 illustrates that the overlap in species identification-collection between pika and packrats of the same fossil unit is from 0 to 2.

Although the packrat midden seems to provide a more comprehensive reconstruction of the local community, it seems likely that the two animals are actually sampling different strata of the local community. The reconstructions based on the packrat midden would tend to imply the paleocommunities were predominantly woodlands or open woodlands with shrubs. The additional data provided by pika

TABLE 5. Modern dung samples of *Ochotona* analyzed for microhistology (presented as percent relative density).

	Ruby Mts., NV	White Mts., CA	Crat. of the Moon, ID	Cedar Breaks, UT	Bear Tooth Mts., WY
GRAZE					
Agropyron sp.	9.15	19.23	11.22	-	-
Bouteloua sp.	-	-	3.54	-	-
Bromus sp.	-	-	-	2.63	-
Carex sp.	5.93	-	-	9.86	15.33
Festuca sp.	12.55	3.39	24.61	-	-
Oryzopsis sp.	24.21	-	-	-	-
Poa sp.	-	-	3.54	14.74	10.25
Stipa sp.	5.93	-	-	-	-
Unkn grass	-	1.65	-	-	-
BROWSE					
Astragalus/ *Oxytropis*	2.89	74.08	-	-	-
Cenothus/ *Cercocarpus*	-	-	11.22	-	-
Compositae	-	-	-	1.30	-
Cowania/ *Pursia*	-	-	3.54	-	-
Draba sp.	2.89	-	-	-	-
Eriogonum sp.	-	-	3.54	-	-
Fraxinus-type	-	-	-	-	68.80
Juniperus sp.	-	1.65	-	4.00	-
Lesquerella sp.	2.89	-	-	-	-
Moss	5.93	-	35.25	1.30	5.62
Pinus sp.	-	-	-	66.17	-
Populus sp.	9.15	-	-	-	-
Ranunculus sp.	12.55	-	-	-	-
Symphoricarpos sp.	-	-	3.54	-	-
Unkn forb	5.93	-	-	-	-

dung analysis would confirm that these woodlands were open and also contained grasses, or that grassland areas occurred near the fossil site.

EXTIRPATION OF PIKA

When did pika move out of their glacial-age habitats in valley bottoms, or near-valley bottoms and mountains ranges, and into their present highly restricted (and possibly ecologically constrained) distribution?

Grayson (1993) discusses the role of pika as a montane mammal of the Pleistocene mountains in the Great Basin in terms of Brown's (1978) insular biogeography theory. The reader is referred to Grayson's work for a thorough discussion of this subject. Pika seem to have become extirpated in low-elevation settings throughout the Great Basin Desert by approximately 7000 yr B.P., according to Grayson (1993). His data consist mainly of skeletal remains recovered from shelters and caves in the Great Basin. Although we agree with his assessment, some of his fossils could be introduced to lower elevation sites as extralimital remains from higher elevations by raptors.

Gatecliff Shelter (Grayson's (1993) main data set) is situated at 2320 m, today in the midst of a pinyon-juniper woodland in the Toquima Range. Pika are found today in the upper reaches of the Toquima Range, the lowest known populations living at 2650 m elevation, approximately 335 m above Gatecliff Shelter. Most of the pika remains were deposited in Gatecliff prior to 5100 yr B.P.; few were found dating less than that age (Grayson, 1983; 1993). Pika were described to be living in a

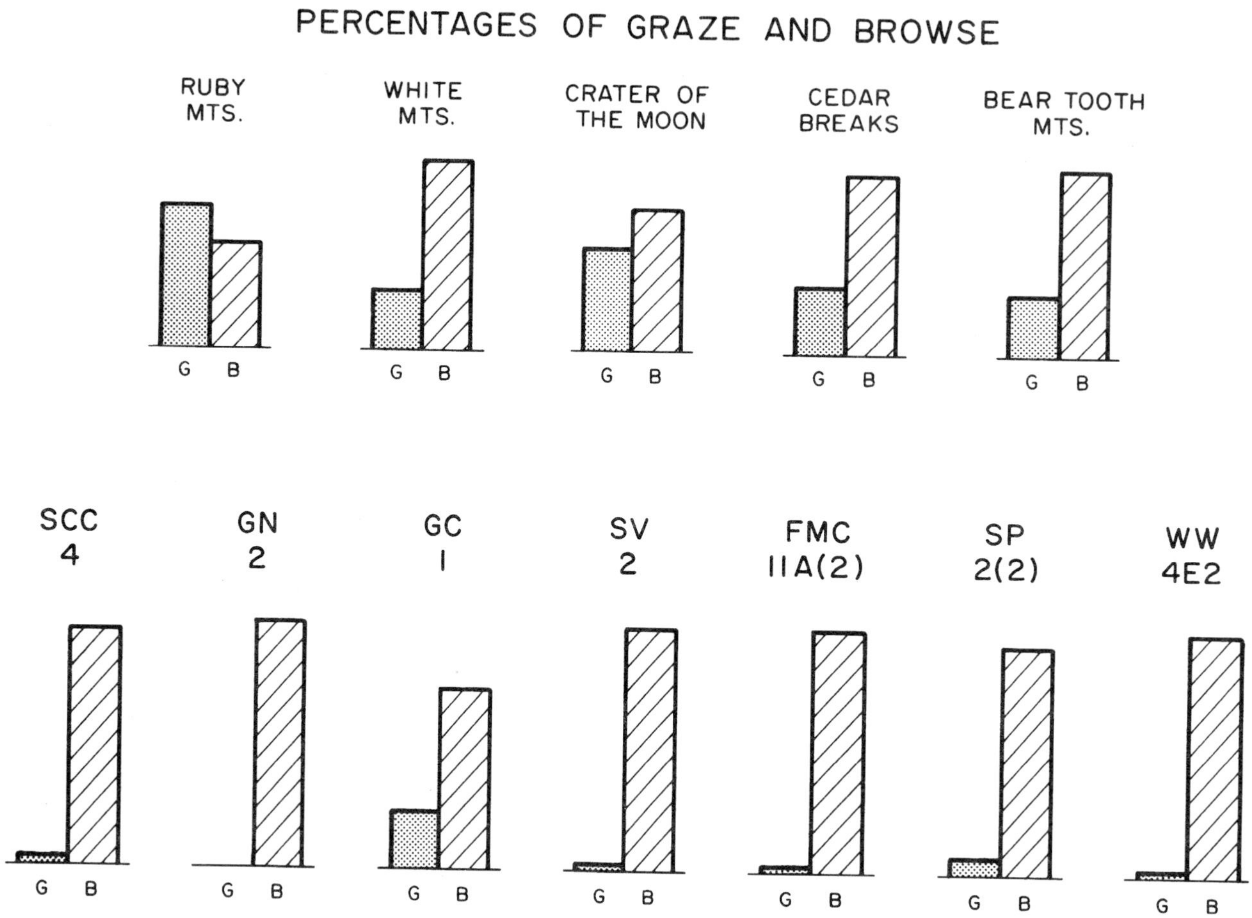

FIG. 9. Histogram of graze (G) and browse (B) species recovered from modern and fossil dung pellets of *Ochotona*.

sagebrush, pinyon-juniper woodland between 6000 and 5100 yr B.P. (Grayson, 1983; Thompson and Kautz, 1983). The decline in numbers of pika fossils in the strata of Gatecliff seem to record the passing of the lower elevational limits of *Ochotona* above the level of the shelter itself. Gatecliff thus seems to show the displacement of *Ochotona* in the mountain range (Grayson, 1993). Grayson (1993) makes a comparison that pika were finally extirpated from low elevations in the Great Basin by about 7000 yr B.P., the same approximate age for the arrival of pinyon pine. This time seems to be a critical ecological point for *Ochotona*.

We have already discussed that pika seem to be heat sensitive and that it would seem that *Ochotona* are not living in many lower elevation habitats due to the present summer high temperature regimes of the northern Mojave and Great Basin deserts. Thompson (1984, 1988) pointed out that summer temperatures of the central Great Basin (e.g., Garrison locality) were about 7° to 9° F cooler in the late glacial than they are today, based on the optimal temperature for photosynthesis in limber pine seedlings. Limber pine and pika were recovered together at the low elevation site of Garrison, just above the high level stand of Lake Bonneville. If Thompson is correct, then it was probably this cooler summer temperatures that allowed pika to inhabit non-talus terrain (as described above) and a plant commu-

TABLE 6. Plant remains from Pleistocene *Ochotona* (O) dung pellets (micohistological material presented as percent relative density in sample) and *Neotoma* (N) middens (macrofossil material abundance scale of Thompson: 1=rare (1 specimen), 2=uncommon (2-20), 3=common (21-100), 4=very common (101-500), and 5=abundant (>500); abundance scale of Spaulding: 1=rare, 2=occasional, 3=common, 4=abundant, and 5=very abundant).
!, nut shell fragments only; in vicinity of site, but more than 50m. All *Ochotona* pellets are from associated *Neotoma* middens.

	Central Great Basin[1]								Northern Mojave[2]					
	12,200 SCC 4		13,400 GN 2		13,600 GC 1		17,300 SV 2		16,400 FMC11A(2)		22,200-30,700 SP2(2)		>44,600 WW 4E	
	O	N	O	N	O	N	O	N	O	N	O	N	O	N
GRAZE														
Agropyron sp.	4.11	-	-	-	4.30	-	-	-	-	-	-	-	-	-
Andropogon sp.	-	-	-	-	-	-	-	-	-	-	2.63	-	1.84	-
Bromus sp.	-	-	-	-	6.64	-	-	-	1.47	-	0.86	-	0.91	-
Festuca sp.	-	-	-	-	14.56	-	-	-	-	-	-	-	-	-
Oryzopsis hymenoides	-	2	-	2	-	-	-	2	-	1	-	-	-	1
Poa sp.	-	-	-	-	2.09	-	-	-	-	-	3.55	-	-	-
Sporobolus sp.	-	-	-	-	-	-	3.18	-	-	-	-	-	-	-
BROWSE: Moss & Conifers														
Moss	4.11	-	-	-	-	-	-	-	-	-	-	-	-	-
Ephedra sp.	-	-	-	-	-	-	-	-	-	-	-	1	1.84	1
E. cf. *viridis*	-	1	-	-	-	-	-	-	-	-	-	-	-	-
Juniperus sp.	-	-	1.68	-	-	-	-	-	-	-	3.55	-	-	-
J. communis	-	2	-	-	-	-	-	4	-	-	-	-	-	-
J. osteosperma	-	-	-	-	-	-	-	-	-	3	-	2	-	5
J. scopulorum	-	3	-	-	-	-	-	2	-	-	-	-	-	-
Picea sp.	89.78	-	-	-	37.38	-	-	-	-	-	-	-	-	-
P. engelmannii	-	4	-	-	-	-	-	-	-	-	-	-	-	-
Pinus sp.	-	-	98.32	-	-	-	-	-	-	-	0.86	-	15.43	-
P. edulis	-	-	-	-	-	-	-	-	-	1	-	-	-	-
P. flexilis	-	-	-	-	-	5	-	5	-	1	-	4	-	-
P. longaeva	-	5	-	5	-	-	-	-	-	-	-	3	-	-
P. monophylla	-	-	-	-	-	-	-	-	-	1!	-	-	-	2
BROWSE: Dicots														
Arabis sp.	-	-	-	-	-	-	-	-	-	-	-	-	-	1
Artemisia ludoviciana	-	-	-	-	-	-	-	-	-	-	-	-	-	1
A. tridentata-type	-	3	-	4	-	4	22.13	-	86.14	4	-	1	-	1
Astragalus/ Oxytropis	-	-	-	-	6.64	-	-	-	-	-	-	-	-	-
Atriplex confertifolia	-	2	-	-	-	-	-	-	-	-	-	2	-	1
Brickellia oblongifolia	-	-	-	-	-	-	-	-	-	1	-	-	-	-
Cercocapus sp.	-	-	-	-	-	-	-	-	-	3	-	-	-	-
C. intricatus	-	-	-	2	-	-	-	-	-	-	-	1	-	2
C. ledifolius	-	-	-	-	-	-	-	-	-	3	-	1	-	-
Cercocarpus/ Ceanothus	-	-	-	-	11.74	-	-	-	1.47	-	12.01	-	-	-
Chamaebatiaria millefolium	-	-	-	2	-	-	-	-	-	1	-	-	-	-
Chenopodium sp.	-	2	-	-	-	3	-	-	-	-	-	-	-	-
Chrysothamnus sp.	-	3	-	3	-	4	-	2	-	-	-	2	-	1
C. nauseosus	-	-	-	-	-	-	-	-	-	1	-	-	-	-
C. cf. *parryi*	-	-	-	-	-	-	-	-	-	1	-	-	-	-
C. viscidiflorus	-	-	-	-	-	-	-	-	-	2	-	-	-	-
Cirsium sp.	-	3	-	-	-	-	-	-	-	-	-	1	-	-
Compositae	-	-	-	-	-	-	-	-	-	-	-	-	1.84	-

Chorizanthe brevicornu-type	-	-	-	-	-	-	-	-	-	1	-	-	-	-
Cryptantha sp.	-	-	-	-	-	-	-	-	-	1	-	1	-	-
Descurainia sp.	-	-	-	-	-	-	-	-	6.42	-	-	-	-	-
Echinocereus sp.	-	2	-	-	-	2	-	-	-	-	-	-	-	-
Eriogonum sp.	-	-	-	-	-	-	-	-	-	2	-	-	-	-
Eurotia sp.	-	-	-	-	-	-	-	-	1.47	-	-	-	-	-
Forsellesia nevadensis	-	3	-	1	-	-	-	-	-	-	-	2	-	1
F. pungens	-	-	-	-	-	-	-	-	-	-	-	-	-	2
Fragaria sp.	-	-	-	-	14.56	-	-	-	-	-	-	-	-	-
Fraxinus anomala	-	-	-	-	-	-	-	-	-	1	-	-	-	-
Gutierrezia sarothrae	-	2	-	-	-	2	-	-	-	-	-	-	-	-
Heterotheca villosa	-	-	-	-	-	-	-	-	-	1	-	-	-	-
Holodiscus microphyllus	-	-	-	-	-	-	-	-	-	1	-	-	-	-
Jamesia americana	-	1	-	-	-	2	-	-	-	-	-	1	-	1
Lepidium sp.	-	2	-	-	-	-	-	1	-	-	-	-	-	-
Lesqurella sp.	-	2	-	-	-	-	-	-	3.03	-	-	-	-	-
Lupinus argenteus-type	-	-	-	-	-	-	-	-	-	1	-	-	-	-
cf.*Machaeranthera* sp.	-	-	-	-	-	-	-	-	-	-	-	-	-	1
cf. *Monardella* sp.	-	-	-	-	-	-	-	-	-	1	-	-	-	-
Oenothera sp.	-	-	-	-	-	-	-	-	-	-	1.73	-	-	-
Opuntia sp.	-	-	-	-	-	-	-	-	-	3	-	-	-	1
O. polycantha	-	-	-	-	-	-	-	-	-	3	-	1	-	-
Phacelia hastata-type	-	-	-	-	-	-	-	-	-	1	-	-	-	-
Philadelphus microphyllus	-	-	-	-	-	-	-	-	-	-	-	1	-	-
Phlox sp.	-	3	-	-	-	2	-	-	-	-	-	-	-	-
Plagiobothrys jonesii	-	-	-	-	-	-	-	-	-	1	-	-	-	-
Potentilla sp.	-	-	-	-	-	-	74.69	-	-	-	-	-	-	-
Prunus sp.	-	2	-	-	-	-	-	-	-	-	-	-	-	-
Rhus sp.	-	-	-	-	-	-	-	-	-	-	-	-	1.84	-
R. trilobata	-	1	-	-	-	-	-	-	-	-	-	-	-	-
Ribes montigenum	-	3	-	-	-	-	-	2	-	-	-	-	-	-
R. cf. *velutinum*	-	-	-	-	-	-	-	-	-	-	-	-	-	1
Rosa sp.	-	2	-	-	-	2	-	-	-	-	-	-	-	-
Salvia sp.	-	-	-	-	-	-	-	-	-	2	-	-	-	-
S. dorrii	-	-	-	-	-	-	-	-	-	2				
Shepherdia sp.	-	2	-	2	-	2	-	2	-	-	-	-	-	-
Sida sp.	-	-	-	-	-	-	-	-	-	-	77.50	-	74.48	-
Symphoricarpos sp.	-	3	-	4	-	3	-	2	-	-	0.86	1	-	3
Tetradymia sp.	-	-	-	-	-	2	-	-	-	2	-	1	-	1
T. canescens-type	-	-	-	-	-	-	-	-	-	2	-	-	-	-
Unkn. forb	2.00	-	-	-	-	-	-	-	-	-	-	-	-	-
Vicia cf. *americana*	-	3	-	-	-	2	-	-	-	-	-	-	-	-

[1]*Neotoma* data from Thompson (1979, 1984) and Thompson and Mead (1982).
[2]*Neotoma* data from Spaulding (1981) and this report.

TABLE 7. Number of identifiable plant taxa from *Ochotona* dung (microhistological remains) and *Neotoma* middens (macrobotanical remains), both preserved in the same packrat midden unit.

	Smith Creek 4	Garrison 2	Granite 1	Stream-view 2	Forty-mile 11	Spires 2	Willow Wash 4
Ochotona	4	2	8	3	2	8	7
Neotoma	25	9	13	9	25	18	18
No. Species in Common[1]	1	1	0	0	1	2	1

[1]Most often the species in common represent a generic identificaiton based on microhisitological analysis (e.g., *Picea*) and a specific identification based on macrohistological analysis (e.g., *Picea engelmannii*) from the same unit (see Table 6).

nity that it avoids today.

The Gatecliff Shelter data may well illustrate when pika were finally extirpated from low-elevation limits of the central Great Basin. When did pika begin to change their Wisconsinan Glacial distributions overall in the Intermountain West? The radiocarbon dates on pika dung-bearing middens in the northern Mojave Desert (Table 1) lead us to propose that pika may been eliminated from at least the lower elevations of the northern Mojave Desert by approximately 14,000 yr B.P. Only one date is younger than that age, and that is from ER 2(11), which has been shown to be a midden stratum that contains a mix of different-age materials (Spaulding, 1985). The relatively early date for the extirpation of pika at lower elevations in the northern Mojave Desert, if true, is consistent with other evidence for increasing temperatures millennia before the close of the Wisconsinan Glacial (Spaulding *et al.,* 1983).

As might be expected at the more northernly sites of the central Great Basin, the radiocarbon data suggest that pika persisted in this region millennia longer than farther south in the northern Mojave Desert (Table 1). Of the nine radiocarbon dated midden assemblages from the central Great Basin containing pika, seven are younger than 14,000 yr B.P. This suggests that there may have been appreciable diachrony in the timing of the local extirpations of this species in the Intermountain West, and that latitudinally mediated changes in the time that climatic thresholds were crossed were likely an important influence on the timing of extirpation - an idea also postulated by Grayson (1993). Pika still occupied low-elevation, isolated inselberg settings within Snake Valley (Garrison, 1640 m elevation) by approximately 12,230 yr B.P. Sometime between approximately 12,000 and 8,000 yr B.P., pika became restricted to and above the lower sections of cooler mountain canyons, such as Smith Creek Canyon. The youngest radiocarbon date on pika at the low-elevation canyon settings (Streamview locality, 1860 m) is 8030 yr B.P. (AMS date directly on *Ochotona* dung). This is in agreement with Grayson's (1993) data indicating that pika distributions at Gatecliff Shelter, Toquima Range were moving upslope, through 2320 m elevation, by approximately 7000 yr B.P. The progression of pika extirpation and range reduction in the northern Mojave and the Great Basin deserts seems to have begun by 14,000 yr B.P. and persisted upslope and northward to the establishment of their modern distribution. When the latter actually happened is not well established, but probably occurred by the end of or during the middle Holocene (7500 to 4500 yr B.P.; Grayson, 1993). This time period in both the northern Mojave and Great Basin deserts was quite warm and/or dry compared to the late glacial and early Holocene (Spaulding, 1991; Thompson, 1990).

SUMMARY

First, the topography of the sites considered here, as well as the taphonomic evidence, shows that the assertion that pika dung pellets in packrat middens were transported to those sites by slope wash (Hafner, 1993) is untenable. In comparing the materials harvested by these two commensals, *Ochotona* and *Neotoma*, we find that glacial-age pika tended to focus more on grasses and forbs, essentially sampling a different stratum of the plant community than did the packrat. There is also evidence for overlap of preferences in some cases. These data suggest no expansion, or shift, of the pika feeding niche comparable to the magnitude of alteration in its habitat range. These habitat alterations occurred on both local and regional scales. Locally, pika were not predictably associated with talus slopes, as they are at present. Although their dung pellets are found in packrat middens recovered from rocky habitats, talus *per se* does not appear to have been a necessary component of its habitat during the last glacial age. Regionally, pika ranged much farther south than they do at present. While this latitudinal displacement mirrors those recorded for many other species during the last glacial age, pika occurred in distinctly different biotic communities than they do at present. The macrofossil assemblages from packrat middens demonstrate that their range extended more than 600 m below subalpine conifer woodland into xeric steppe shrub-juniper woodland. Therefore, pika remains cannot be used as indicators of boreal habitats in the paleontologic record. We suggest that this lack of habitat fidelity from glacial to interglacial times may well extend to other species.

ACKNOWLEDGMENTS

We thank Dr. Jerry H. Theis and his colleagues at the School of Medicine, University of California, Davis, for their work on the nuclear magnetic resonance. We appreciate Emilee M. Mead (University of Arizona) who helped on the initial field collecting of pika dung pellets and provided the line drawings. Donald K. Grayson (University of Washington) has shown enthusiasm over the many years of our research in the Great Basin. We thank him for his continued support of our research.

REFERENCES

Beatly, J. C. (1976). Vascular plants of the Nevada Test Site and central southern Nevada. *National Technical Information Service Report* TID-26881.

Bell, C. J. (1990). "A Rancholabrean Mammalian Fauna from Cathedral Cave, White Pine County, Nevada." Unpublished M.S. thesis, Northern Arizona University, Flagstaff.

Betancourt, J., Martin, P. S., and Van Devender, T. R., Eds. (1990). "Packrat Middens the Last 40,000 Years of Biotic Change." University of Arizona Press, Tucson.

Bleich, V. C., and Schwartz, O. A. (1975). Observations on the home range of the desert woodrat, *Neotoma lepida intermedia*. *Journal of Mammalogy* **56**, 518-519.

Brown, J. H. (1978). The theory of insular biogeography and the distribution of boreal birds and mammals. *In* "Intermountain Biogeography: a Symposium" (K. T. Harper and J. L. Reveal, Eds.), pp. 209-227. *Great Basin Naturalist Memoirs* **2**, 1-268.

Churcher, C. S., and Dods, R. R. (1979). *Ochotona* and other vertebrates of possible Illinoian age from Kelso Cave, Halton County, Ontario. *Canadian Journal of Earth Sciences* **16**, 1613-1620.

Corbet, G. B. (1978). "The Mammals of the Palaearctic Region: a Taxonomic Review." British Museum and Cornell University Press, London and Ithaca.

Erbajeva, M. A. (1988). ["Pika of the Cenozoic Age"] Nauka, Moscow-Leningrad. [In Russian].

Grayson, D. K. (1983). The paleontology of Gatecliff Shelter: small mammals. *In* "The Archaeology of Monitor Valley. 2. Gatecliff Shelter" (D. H. Thomas, Ed.), pp. 98-126. American Museum of Natural History Anthropological Papers 59(1).

Grayson, D. K. (1993). "The Desert's Past. A Natural Prehistory of the Great Basin." Smithsonian Institution Press, Washington.

Guthrie, R. D., and Matthews, J. V. (1971). The Cape Deceit fauna-early Pleistocene mammalian assemblage from the Alaskan Arctic. *Quaternary Research* **1**, 474-510.

Hafner, D. J. (1993). North American pika (*Ochotona princeps*) as a late Quaternary biogeographic indicator. *Quaternary Research* **39**, 373-380.

Hall, E. R. (1946). "Mammals of Nevada." University of California Press, Berkeley.

Hall, E. R. (1981). "Mammals of North America." John Wiley and Sons, New York.

Huntly, N. J. (1987). Influence of refuging consumers (pikas: *Ochotona princeps*) on subalpine meadow vegetation. *Ecology* **68**, 274-283.

Huntly, N. J., Smith, A. T., and Ivins, B. L. (1986). Foraging behavior of a refuging herbivore, the pika (*Ochotona princeps*: Lagomorpha), with comparisons of grazing and caching. *Journal of Mammalogy* **67**, 139-148.

Mead, E. M., and Mead, J. I. (1989). Snake Creek Burial Cave and a review of the Quaternary mustelids of the Great Basin. *Great Basin Naturalist* **49**, 143-154.

Mead, J. I. (1987). Quaternary records of pika, *Ochotona*, in North America. *Boreas* **16**, 165-171.

Mead, J. I., and Murray, L. K. (1991). Late Pleistocene vertebrates from the Potosi Mountain packrat midden, Spring Range, Nevada. *San Bernardino County Museum Association, Special Publication, Mojave Desert Quaternary Research Center Symposium* **91**, 124-126.

Mead, J. I., Bell, C. J., and Murray, L. K. (1992). *Mictomys borealis* (northern bog lemming) and the Wisconsin paleoecology of the east-central Great Basin. *Quaternary Research* **37**, 229-238.

Mehringer, P. J. (1967). Pollen analysis of the Tule springs area, Nevada . *Nevada State Museum Anthropological Papers* **13**, 129-200.

Millar, J. S., and Zwickel, F. C. (1972). Characteristics and ecological significance of hay piles of pikas. *Mammalia* **36**, 657-667.

Mitchell, R. M. (1981). The *Ochotona* (Lagomorpha: Ochotonidae) of Asia. *In* "Geological and Ecological Studies of Qinghai-Xizang Plateau" (Science Press, Eds.), pp. 1031-1038. Proceedings of Symposium on Qinghai-Xizang (Tibet) Plateau. Science Press, Beijing.

Orr, R. T. (1977). "The Little-Known Pika." Macmillan Publishing Co., New York.

Smith, A. T. (1974). The distribution and dispersal of pikas: influences of behavior and climate. *Ecology* **55**, 1368-1376.

Smith, A. T., and Weston, M. L. (1990). Ochotona princeps. *Mammalian Species* **352**, 1-8.

Smith, A. T., Formozov, N. A., Hoffmann, R. S., Changlin, Z., and Erbajeva, M. A. (1990). The pikas. *In* "Rabbits, Hares and Pikas, Status Survey and Conservation Action Plan" (J. A. Chapman and J. E. C. Flux, Eds.), pp. 14-60. IUCN/SSC Lagomorph Specialist Group.

Spaulding, W. G. (1981). "The Late Quaternary Vegetation of a Southern Nevada Mountain Range." Unpublished Ph.D. dissertation, University of Arizona, Tucson.

Spaulding, W. G. (1985). Vegetation and climates of the last 45,000 years in the vicinity of the Nevada Test site, south-central Nevada. *US Geological Survey Professional Paper* **1329**.

Spaulding, W. G. (1990). Vegetational and climatic development of the Mojave Desert: the last glacial maximum to the present. *In* "Packrat Middens the Last 40,000 Years of Biotic Change" (J. L. Betancourt, T. R. Van Devender, and P. S. Martin, Eds.), pp.166-199. University of Arizona Press, Tucson.

Spaulding, W. G. (1991). A middle Holocene vegetation record from the Mojave Desert of North America and its paleoclimatic significance. *Quaternary Research* **35**, 427-437.

Spaulding, W. G., Betancourt, J. L., Croft, L. K., and Cole, K. L. (1990). Packrat middens: their composition and methods of analysis. *In* "Packrat Middens the Last 40,000 Years of Biotic Change" (J. L. Betancourt, T. R. Van Devender, and P. S. Martin, Eds.), pp. 59-84. University of Arizona Press, Tucson.

Spaulding, W. G., Leopold, E. B., and Van Devender, T. R. (1983). Late Wisconsin paleoecology of the American Southwest. *In* "Late-Quaternary Environments of the United States." Volume 1, "The Late Pleistocene" (S. C. Porter, Ed.), pp. 259-293. University of Minnesota Press, Minneapolis.

Theis, J. H., de Ropp, J. S., Schwab, R. G., Banks, J., and Levine, K. (1988). Nuclear magnetic resonance to differentiate bear, pig and cow bile for forensic investigations. *Wildlife Society Bulletin* **16**, 430-433.

Thompson, R. S. (1979). Late Pleistocene and Holocene packrat middens from Smith Creek Canyon, White Pine County, Nevada. *Nevada State Museum Anthropological Papers* **17**, 362-380.

Thompson, R. S. (1984). "Late Pleistocene and Holocene Environments in the Great Basin." Unpublished Ph.D. dissertation, University of Arizona, Tucson.

Thompson, R. S. (1988). Vegetation dynamics in the western United States: modes of response to climatic fluctuations. *In* "Vegetation History" (B. Huntley and T. Webb, Eds.), pp. 415-458. Kluwer, Dordrecht, The Netherlands.

Thompson, R. S. (1990). Late Quaternary vegetation and climate in the Great Basin. *In* "Packrat Middens the Last 40,000 Years of Biotic Change" (J. L. Betancourt, T. R. Van Devender, and P. S. Martin, Eds.), pp. 200-239. University of Arizona Press, Tucson.

Thompson, R. S., and Kautz, R. R. (1983). Paleobotany of Gatecliff Shelter: pollen analysis. *In* "The Archaeology of Monitor Valley. 2. Gatecliff Shelter" (D. H. Thomas, Ed.), pp. 136-151. American Museum of Natural History Anthropological Papers **59**(1).

Thompson, R. S., and Mead, J. I. (1982). Late Quaternary environments and biogeography in the Great Basin. *Quaternary Research* **17**, 39-55.

Van Devender, T. R., Martin, P. S., Thompson, R. S., Cole, K. L., Jull, A. J. T., Long, A., Toolin, L. J., and Donahue, D. J. (1985). Fossil packrat middens and the tandem accelerator mass spectrometer. *Nature* **317**, 610-613.

Van Devender, T.R., and Spaulding, W.G. (1979). Development of vegetation and climate in the southwestern United States. *Science* **204**, 701-710.

Van Devender, T. R., Thompson, R. S., and Betancourt, J. L. (1987). Vegetation history in the Southwest: the nature and timing of the late Wisconsin-Holocene transition. *In* "North America and Adjacent Oceans During the Last Deglaciation" (W. F. Ruddiman and H. E. Wright, Eds.), pp. 323-352. Geological Society of America, K-3, Boulder.

Weston, M. L. (1981). The *Ochotona alpina* complex: a statistical re-evaluation. *In* "Proceedings of the World Lagomorph Conference" (K. Myers, and C. D. MacInnes, Eds.), pp. 73-89. University of Guelph, Ontario.